"The Yeast Connection Handbook *is Dr. William Crook's most recent book on the manifold health problems that may be related to yeast infections. It will be an indispensable guide to diagnosis and treatment for doctors and their patients."*

—Cory SerVaas, M.D., Editor,
The Saturday Evening Post

"A wonderful book! Interesting and timely. Comprehensive, yet concise. This successor to The Yeast Connection *will help countless people of all ages and both sexes."*

—Martie Whittekin, CCN, Past President,
National Nutritional Foods Association,
Plano, Texas

"The Yeast Connection Handbook *includes* all *of the new information found in* The Yeast Connection and the Woman *in a readable form. Added features include: Easier to follow diets; a question and answer chapter; and new material about yeast-connected disorders in children."*

—Elmer Cranton, M.D., Past President,
The American College for Advancement in Medicine and Past President, The American Holistic Medical Association,
Yelm, Washington

"The shelf of medical books at my desk now has a marvelous addition: a resource for people who have sought help tirelessly, but unsuccessfully, for troubling health problems. In Dr. Crook, they will find a friend and professional who makes sense of their symptoms and provides a blueprint for long-sought good health."

—Sue MacDonald, Health Writer,
Cincinnati Enquirer

"Many people today suffer from chronic unwellness. This up-to-date book presents new information on a topic whose time has come."

—George Miller, M.D., Fellow,
American College of Obstetricians and Gynecologists, Lewisberg, Pennsylvania

Are Your Health Problems Yeast Connected?

If your answer is "yes" to any question, circle the number in the right hand column. When you've completed the questionnaire, add up the points you've circled. Your score will help you determine the possibility (or probability) that your health problems are yeast connected.

	Yes	No	Score
1. Have you taken repeated or prolonged courses of antibacterial drugs?	☐	☐	4
2. Have you been bothered by recurrent vaginal, prostate or urinary infections?	☐	☐	3
3. Do you feel "sick all over," yet the cause hasn't been found?	☐	☐	2
4. Are you bothered by hormone disturbances, including PMS, menstrual irregularities, sexual dysfunction, sugar craving, low body temperature or fatigue?	☐	☐	2
5. Are you unusually sensitive to tobacco smoke, perfumes, colognes and other chemical odors?	☐	☐	2
6. Are you bothered by memory or concentration problems? Do you sometimes feel "spaced out"?	☐	☐	2
7. Have you taken prolonged courses of prednisone or other steroids; or have you taken "the pill" for more than 3 years?	☐	☐	2
8. Do some foods disagree with you or trigger your symptoms?	☐	☐	1
9. Do you suffer with constipation, diarrhea, bloating or abdominal pain?	☐	☐	1
10. Does your skin itch, tingle or burn; or is it unusually dry; or are you bothered by rashes?	☐	☐	1
TOTAL SCORE			

Scoring for women: If your score is 9 or more, your health problems are probably yeast connected. If your score is 12 or more, your health problems are almost certainly yeast connected.

Scoring for men: If your score is 7 or more, your health problems are probably yeast connected. If your score is 10 or more, your health problems are almost certainly yeast connected.

William G. Crook, M.D.

PROFESSIONAL BOOKS,INC.
Jackson, Tennessee

· DISCLAIMER ·

This book describes relationships that have been observed between the common yeast germ *Candida albicans* and health problems that affect people of all ages and both sexes—especially premenopausal women. *I have written it to serve only as a general informational guide and reference source for both professionals and nonprofessionals.*

For obvious reasons I cannot assume the medical or legal responsibility of having the contents of this book considered as a prescription for anyone.

Treatment of health disorders, including those which appear to be yeast connected, must be supervised by a physician or other licensed health professional. Accordingly, you and the professional who examines and treats you must take the responsibility for the uses made of this book.

▪ ▪ ▪

Published by Professional Books, Inc., Box 3246, Jackson, Tennessee 38303.

Cover art, book design and typography by ProtoType Graphics, Inc., Nashville, Tennessee.

Library of Congress Catalog Card Number: 95–067424

ISBN: 0-933478-24-0

Manufactured in the United States of America
8 9 10 11 12 13 14 15 — 05 04 03 02 01 00

Table of Contents

DEDICATION

To the millions of people
with yeast-related
health problems.

ACKNOWLEDGMENTS

I want to again acknowledge my special indebtedness to C. Orian Truss, M.D., of Birmingham, Alabama. This brilliant and courageous internist first noted the relationship of superficial yeast infections to health problems that affect people of all ages and both sexes. Thanks also are due to hundreds of people, including professionals and nonprofessionals, who have shared their knowledge with me. As they come across their ideas in this book, they will know how grateful I am for their help and consultation.

As with my other publications, I'm grateful to John Adams and the entire staff of ProtoType Graphics, Inc., Nashville, Tennessee, for their skillful production services.

Special words of appreciation are again due to my daughter, Cynthia, for her delightful illustrations and to Gregg Bender for his illustration of the antibiotic cascade.

I'm grateful also to Janet Gregory, who again typed the manuscript; to Mary Reed, who edited it; to my daughter, Elizabeth, who serves as my business consultant; and to Brenda Harris, Jan Torre and Georgia Deaton, who helped me put the book together.

Foreword

I didn't have a clue about the profound influence yeast-related illness would have on the nature of my medical practice. Then, as a young allergy fellow-in training, I was confronted with a patient with a compelling history. I had just finished treating her for a complex illness involving multiple food and chemical sensitivities—unsuccessfully I might add.

Somewhat later, she made a fateful telephone call to inform me that she was now doing "just fine" . . . Not by what I had done for her but because of what she had learned about her yeast-related illness from Dr. C. Orian Truss of Birmingham, Alabama. So began a personal odyssey for me, and for many of my patients, as I sought to unravel the complexities of yeast-related illness with the help of knowledgeable physicians such as Dr. Truss and Dr. William Crook.

Time and time again since then, patients have presented themselves to my office for treatment saying that by following the advice in Dr. Crook's book, *The Yeast Connection,* before seeing me, they began to feel well again, often after years of inexplicable illness.

As patients' knowledge increased, I have observed a simultaneous change in their needs. For example, whereas ten years ago, patients had almost no information available on yeast-related illness, now many patients seem "deluged" with advice . . . Some is helpful, yet, some is not.

In my 17+ years of treating patients with candida-related illness, I'm continually impressed with recurring (but time consuming) issues that the physician treating this illness must come to grips with. The first issue is the regularity with which patients ask questions that must be answered.

These include: Do I have to stay on a special diet and antifungal medications forever? What about artificial sweeteners? What will happen if I need to take an antibiotic for a serious infection? etc. One of the great strengths of this new book is that Dr. Crook has had the foresight, experience (and yes, courage) to address these difficult questions.

Secondly, I'm always impressed that yeast-related illness is rarely a problem in-and-of itself for a patient, but, usually is associated with other illnesses, such as multiple food and/or chemical sensitivities; mold allergy; thyroid dysfunction and nutrient deficiencies. If patients are to succeed in overcoming their illnesses, they must be empowered with knowledge regarding these issues, and treated by their physicians in a comprehensive fashion, I know of no other book that will give patients with yeast-related issues a better "reference point" for gaining perspective on the illness than this book; it will certainly make my job as a practitioner immeasurably easier.

In short, it is truly a rare medical book that is indispensable to both patients and physicians alike. *The Yeast Connection Handbook* is such a book. Dr. Crook's literary efforts have kept pace with patients' increasing and changing questions related to yeast-related illness. I can sum up my opinion of this book in two simple words: REQUIRED READING.

—George F. Kroker, M.D., Fellow,
American College of Allergy,
Asthma and Immunology, LaCrosse,
Wisconsin

Why This Book Was Written

This book was written to bring you up-to-date information on yeast-related health problems and easy-to-follow steps you can take to overcome them.

My first yeast book, *The Yeast Connection,* was published in hard cover in 1983, with a second edition in 1984. Then, it was expanded and updated in a soft cover third edition and published in 1986. Since that time, only minor changes have been made.

During the late 1980s and early and mid-90s, new scientific observations and medical reports provided support for the yeast connection to PMS, psoriasis, asthma, multiple sclerosis, chronic fatigue syndrome, endometriosis, vulvodynia (burning vulva), interstitial cystitis and autism. There was also new information about:

- The role of simple sugars in promoting yeast overgrowth in the digestive tract.
- Prescription antiyeast medications, including Diflucan and Sporanox.
- Nonprescription antiyeast agents, including probiotics, garlic and citrus seed extracts.

There was also an "explosion" of knowledge about:

- Supplemental vitamins, minerals and essential fatty acids.
- Free radicals and their role in contributing to chronic health disorders and the importance of antioxidants in controlling them.

- Ginkgo biloba and CoQ_{10} and other nonprescription remedies that contribute to better health.
- The importance of vegetables and fruits and their role in preventing cancer and other chronic health disorders.
- The mind/body connection.
- Food and chemical sensitivities.
- Thyroid and adrenal hormones.

I gathered all this new material together and included it in my 750-page book, *The Yeast Connection and the Woman,* published in the late spring of 1995.* In his introduction, Philip K. Nelson, M.D., of Sarasota, Florida, said:

> The Yeast Connection and the Woman *is superb. As with his previous publications, this book is well organized and written— immensely readable... up-to-the-minute, factual and practical... this is a book that will be useful to any reader*... The suggestions are practical, affordable and scientifically based. Sick and well alike would benefit if each one were to adopt the healthy lifestyle proposed.

I was, of course, delighted with Dr. Nelson's introduction. Moreover my new book received praise from many people, including readers of *The Yeast Connection* who were looking for new information. In October 1995, I had dinner with Terry Oldham and Kay Longworth whose experiences with interstitial cystitis are described in *The Yeast Connection and the Woman.* In discussing it, they said:

> Dr. Crook, your new book is wonderful! Comprehensive and informative. BUT, many people with yeast-related health problems are overwhelmed by a book of this size. Why don't you write a condensed version with simple, easy-to-follow instructions. Then, after reading it, people who need more information can read your big book . . .

And that, my friends, is why this book was written.

*A revised fourth edition was published in 1998.

CHAPTER · ONE

The Yeast Connection: An Overview

■ ■ ■

Are Your Health Problems Yeast Connected?

If you . . . feel "sick all over"

- have taken many antibiotic drugs
- are bothered by fatigue, headache or depression
- often feel spaced out
- are bothered by muscle aches and digestive problems

- crave sugar
- are unusually sensitive to tobacco, perfume and other chemicals
- are bothered by food sensitivities
- have sought in vain for help from many different physicians

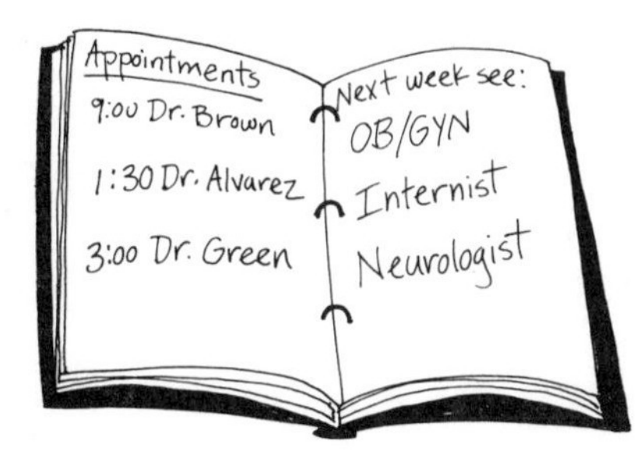

your health problems are probably yeast connected.

Women between 20 and 50 are especially apt to develop yeast-related problems. Common symptoms* include:

- Recurrent vaginal yeast infections
- PMS
- Recurrent urinary tract infections
- Sexual dysfunction
- Dyspareunia (pain on intercourse)

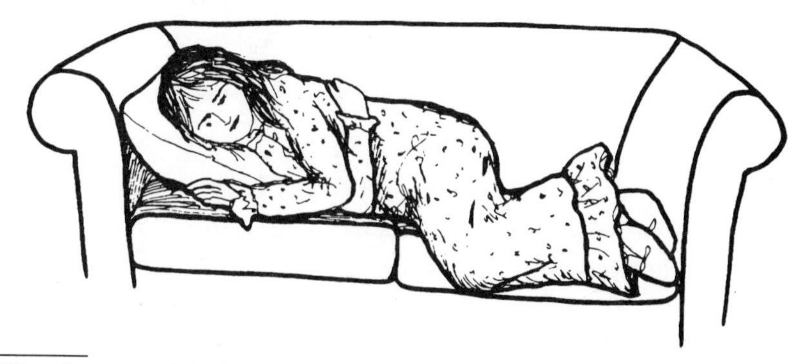

*See also pages 23–27.

Other problems that affect many women also may be yeast-connected, including:

- Vulvodynia (burning vulva)
- Endometriosis
- Interstitial cystitis
- Infertility

Men and children also develop yeast-related problems, especially those who take many antibiotic drugs or steroids.

Common Symptoms in Male Adults

- Fatigue
- Headache
- Digestive symptoms
- Muscle and joint pains

- Depression
- Chemical sensitivities
- Food sensitivities
- Sugar craving
- Memory loss
- Sexual dysfunction

Common Symptoms in Infants and Young Children

- Constant colds
- Irritability
- Sleep problems
- Digestive problems

- Attention deficits
- Skin rashes
- Ear problems
- Hyperactivity

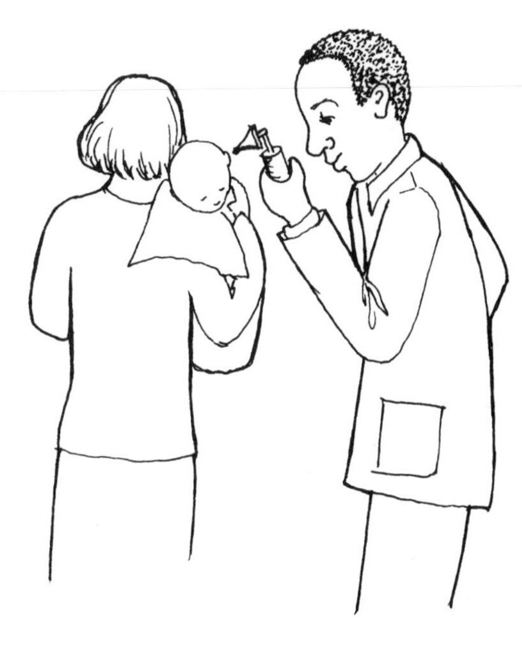

Common Symptoms in Older Children and Teenagers

- Fatigue
- Poor school performance
- Depression
- Food and chemical sensitivities
- Irritability
- Headache

Other complaints and illnesses in adults of both sexes that sometimes may be yeast-related, include:

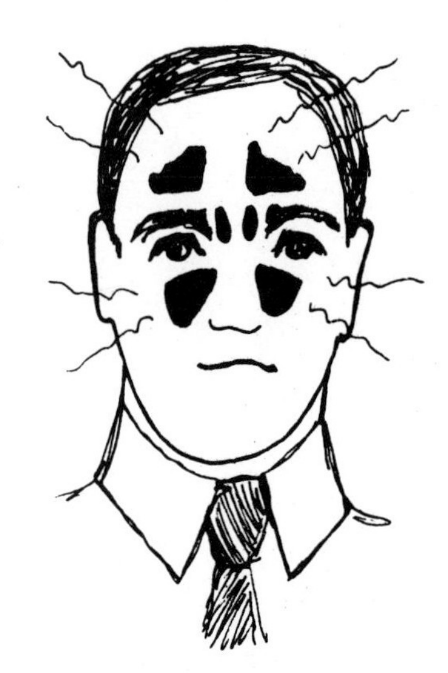

- Numbness
- Crohn's disease
- Scleroderma
- Myasthenia gravis
- Eczema
- Lupus erythematosus
- Psoriasis
- Multiple sclerosis
- Sinusitis*
- Rheumatoid arthritis
- Tingling
- Acne
- Asthma
- Chronic hives

I'm not saying that the common yeast, *Candida albicans,* is *the* cause of all of these problems. Yet, candida may be one of the causes—even a major cause—of these and other health problems.

Yeasts—What They Are and How They Make You Sick

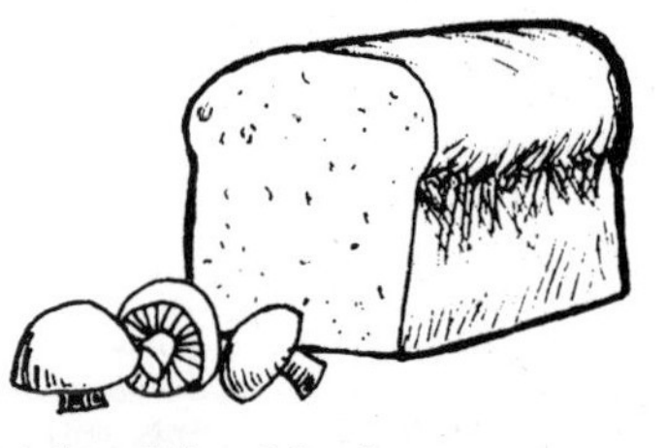

Yeasts are single-cell living organisms that are neither animal nor vegetable. They live on the surfaces of all living things, including fruits, vegetables, grains and your skin. They're a part of the "microflora," which contributes in various ways to the health of its host.

Yeast, itself, is nutritious and small amounts of yeasts give bread its good yeasty taste. Yeast is a kind of fungus. Mildew,

*J. U. Ponikau and colleagues, in an article, "The Diagnosis and Incidence of Allergic Fungal Sinusitis," found positive fungal cultures of nasal secretions in 202 (96%) of 210 consecutive chronic rhinosinusitis patients. (Mayo Clinic Proceedings, 1999; 74:877-884).)

mold, mushrooms, monilia and candida are all names that are used to describe different types of yeast.

One family of yeasts, *Candida albicans,* normally lives on the inner warm creases and crevices of your digestive tract and vagina.

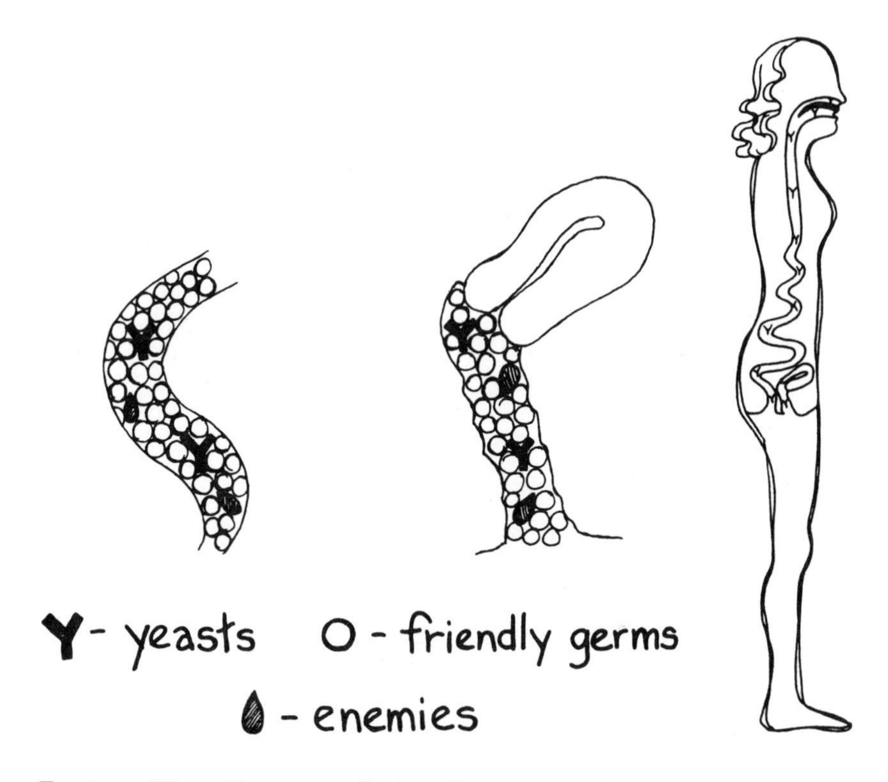

Factors That Promote Yeast Overgrowth

Prescription Medications

Yeast-related health problems are especially apt to trouble you if you've taken repeated or prolonged courses of amoxicillin, ampicillin, Ceclor, Keflex, the tetracyclines and/or other broad-spectrum antibiotics during infancy, childhood and adolescence, or since you've become an adult.

Here's how that happens. As I've said, the common yeast, *Candida albicans,* normally lives in your body, especially in your intestines and vagina. When your immune system is strong, candida yeasts cause no problems. But, when you take broad-spectrum antibiotics for such conditions as acne, respiratory infections or cystitis, these drugs knock out friendly germs while they're knocking out enemies.

Candida yeasts aren't affected by antibiotics so they multiply and raise large families.

These yeasts put out toxins

that weaken the immune system.

So, you may experience repeated infections. Each infection is treated with another round of antibiotics, encouraging further yeast overgrowth, and a vicious cycle develops.

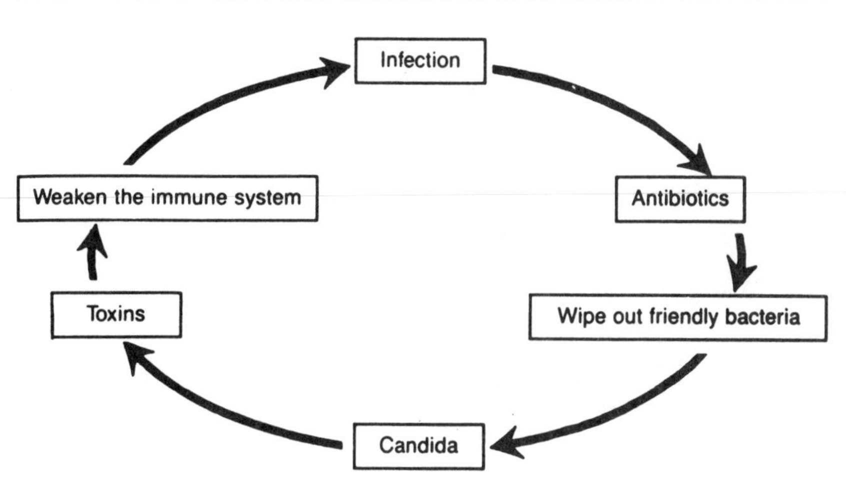

Sugar and Other Simple Carbohydrates

A diet rich in sugar and other simple carbohydrates promotes yeast overgrowth.

Other Factors

Yeast overgrowth also can be caused by other factors, including:

- Hormonal changes associated with the normal menstrual cycle.
- Birth control pills.
- Pregnancy.
- Steroids, taken by pill, injection or inhalation.
- Genital irritations and abrasions.
- Re-infection from your sexual partner.
- Diabetes.

The yeast infection in your genital area also may be related to wearing jockey shorts or nylon underwear.

The "Yeast Connection"

The "yeast connection" is a term to indicate the relationship of superficial yeast infections in your digestive tract (or vagina) to fatigue, headache, depression, PMS, irritability and other symptoms that can make you feel "sick all over."

Other causes must also be considered. If you're like most people with a candida-related health problem, you resemble an overburdened camel. To regain your health, to look good, feel good and enjoy life, you'll need to unload many "bundles of straw." This may take months—even a year or two—but then your camel will be off and running.

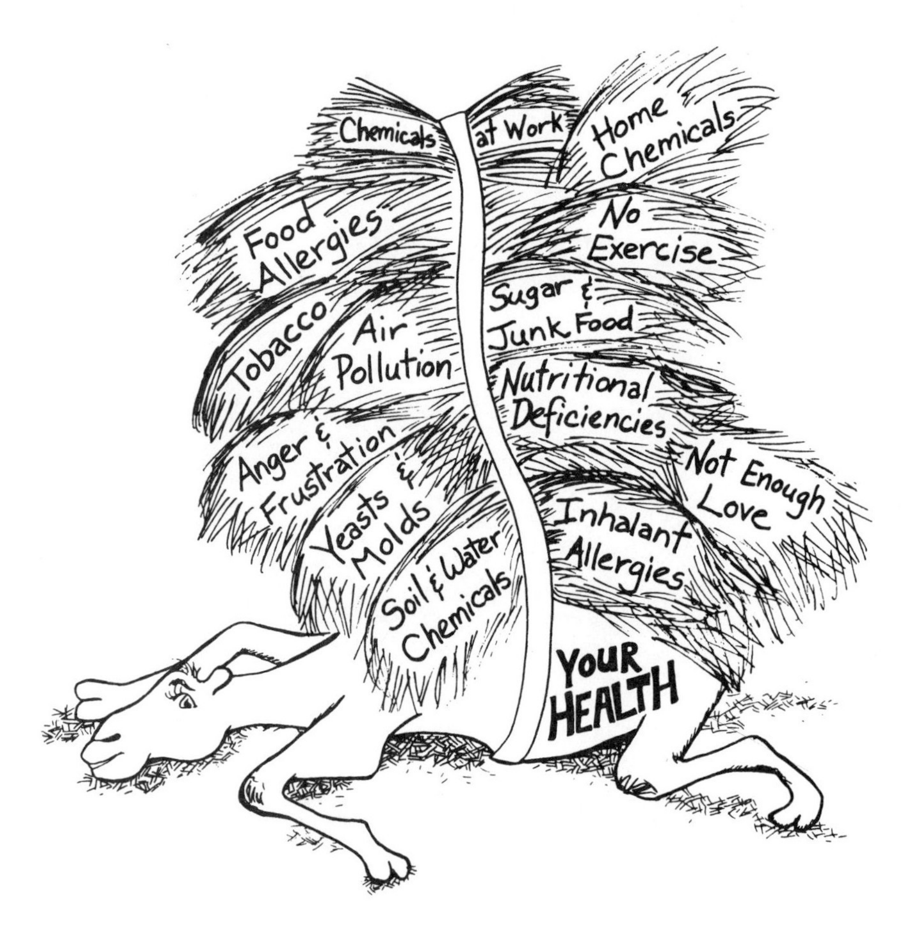

How Superficial Yeast Infections Cause Symptoms in Distant Parts of Your Body

A yeast infection in one part of your body can cause symptoms elsewhere in several different ways:

Immune System Disturbances

Studies by Japanese researchers at the University of Tokyo show that *Candida albicans* puts out high and low molecular weight toxins that can weaken your immune system:[1]

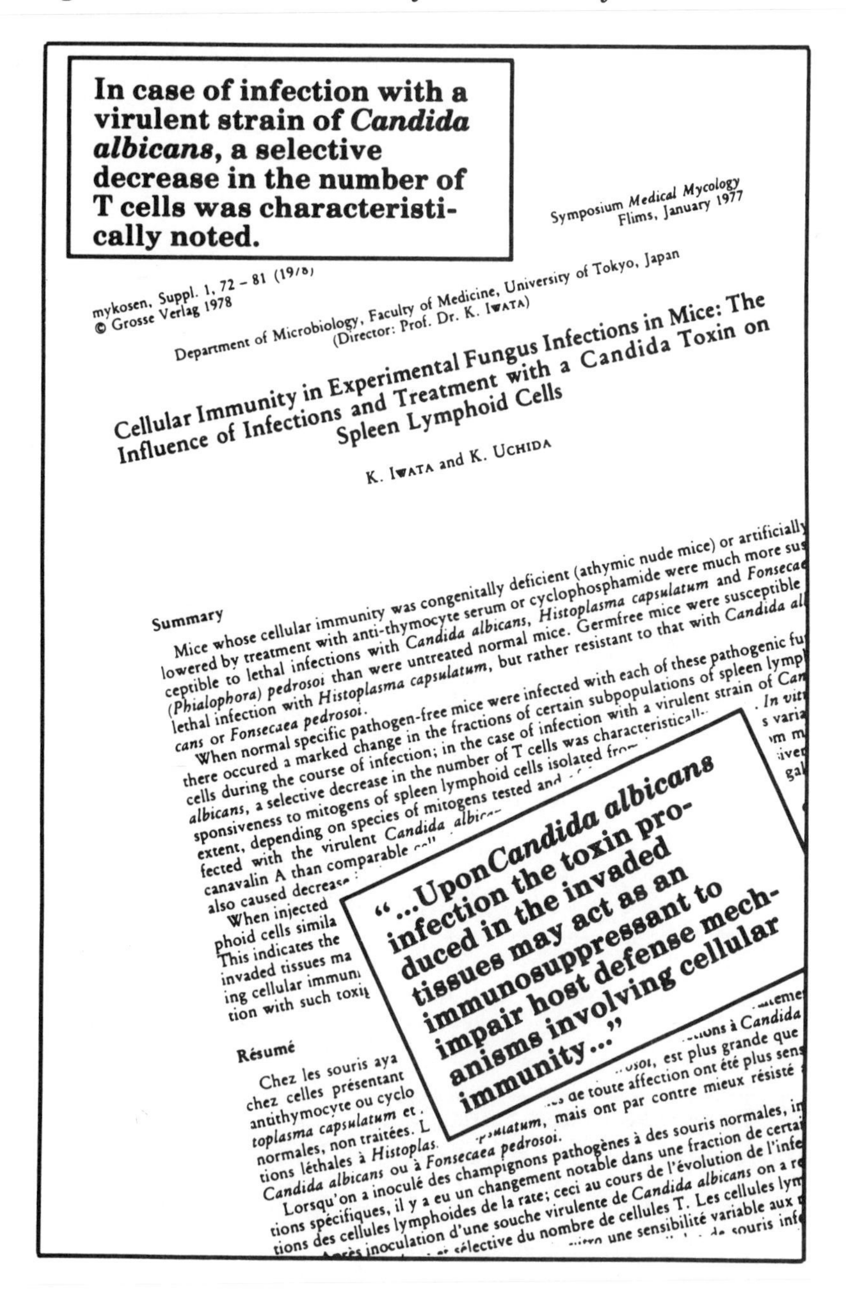

In case of infection with a virulent strain of *Candida albicans*, a selective decrease in the number of T cells was characteristically noted.

Symposium *Medical Mycology*
Flims, January 1977

mykosen, Suppl. 1, 72 – 81 (1978)
© Grosse Verlag 1978

Department of Microbiology, Faculty of Medicine, University of Tokyo, Japan
(Director: Prof. Dr. K. Iwata)

Cellular Immunity in Experimental Fungus Infections in Mice: The Influence of Infections and Treatment with a Candida Toxin on Spleen Lymphoid Cells

K. Iwata and K. Uchida

Summary

Mice whose cellular immunity was congenitally deficient (athymic nude mice) or artificially lowered by treatment with anti-thymocyte serum or cyclophosphamide were much more susceptible to lethal infections with *Candida albicans*, *Histoplasma capsulatum* and *Fonsecaea (Phialophora) pedrosoi* than were untreated normal mice. Germfree mice were susceptible to lethal infection with *Histoplasma capsulatum*, but rather resistant to that with *Candida albicans* or *Fonsecaea pedrosoi*.

When normal specific pathogen-free mice were infected with each of these pathogenic fungi, there occurred a marked change in the fractions of certain subpopulations of spleen lymphoid cells during the course of infection; in the case of infection with a virulent strain of *Candida albicans*, a selective decrease in the number of T cells was characteristically ... *In vitro* ... sponsiveness to mitogens of spleen lymphoid cells isolated from ... extent, depending on species of mitogens tested and ... fected with the virulent *Candida albicans* ... canavalin A than comparable cell ... also caused decrease ...

When injected ... phoid cells simila... This indicates the ... invaded tissues ma... ing cellular immun... tion with such toxi...

"...Upon *Candida albicans* infection the toxin produced in the invaded tissues may act as an immunosuppressant to impair host defense mechanisms involving cellular immunity..."

Résumé

Chez les souris aya... chez celles présentant ... antithymocyte ou cyclo... *toplasma capsulatum* et ... normales, non traitées. L... tions léthales à *Histoplas*... *Candida albicans* ou à *Fonsecaea pedrosoi*.

Lorsqu'on a inoculé des champignons pathogènes à des souris normales, ... tions spécifiques, il y a eu un changement notable dans une fraction de certai... tions des cellules lymphoides de la rate; ceci au cours de l'évolution de l'infe... Après inoculation d'une souche virulente de *Candida albicans* on a r... ...sélective du nombre de cellules T. Les cellules lym... ...vitro une sensibilité variable aux ... de souris inf...

Other researchers have also noted the relationship of intestinal and vaginal yeast infections to immunological problems.[2,3]

Absorption of Food Antigens—and Toxins

Based on clinical and research studies by many different observers, candida overgrowth in your intestine may create what has been called a "leaky gut." Toxins and food allergens* may then pass through this membrane and go to other parts of your body, making you feel "sick all over."

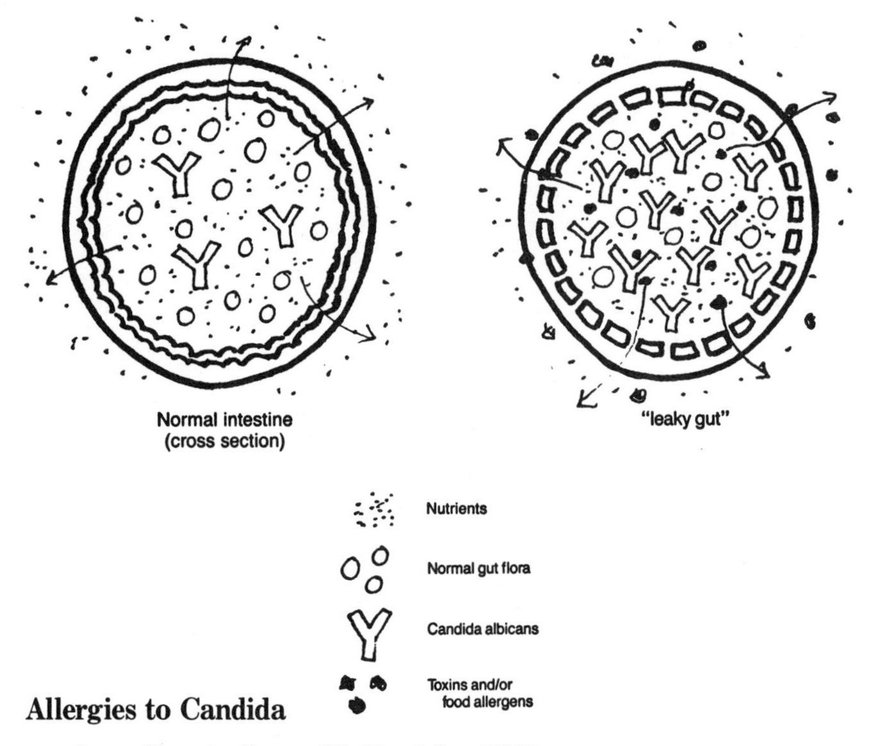

Normal intestine (cross section)

"leaky gut"

Allergies to Candida

According to James H. Brodsky, M.D.:

> There is much evidence to suggest that *C. albicans* is one of the most allergenic microbes. Both immediate and delayed hypersensitivity reactions to candida are very common in the adult population.[4]

---

*The role of probiotics in the management of food allergies was reported in the *Journal of Allergy and Clinical Immunology* (1997; 99: 179–85). See also page 77.

Still Other Factors Often Play a Role in Making You Sick

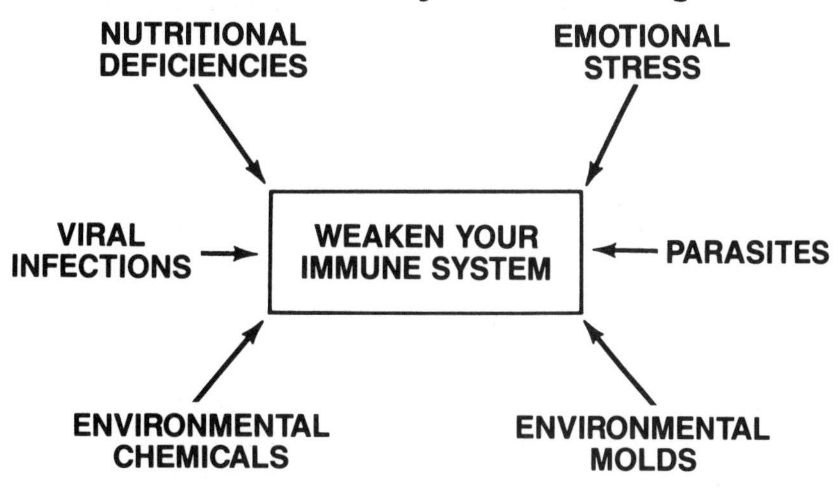

Many of your yeast problems, including fatigue, headache, depression, PMS and sexual dysfunction, develop because your immune system, your endocrine system and your brain are intimately related.

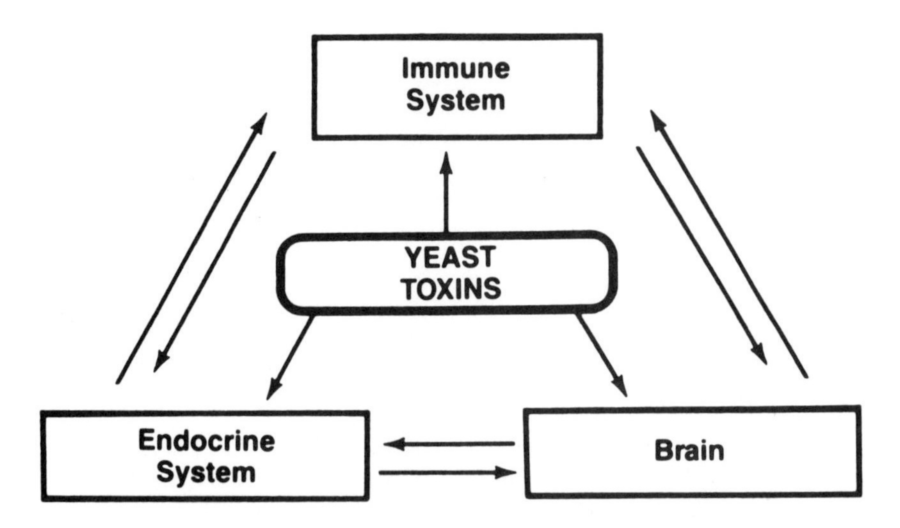

And (although we sometimes forget it) ***every*** part of your body is connected to ***every*** other part.

To summarize: Many different factors play a part in making you sick. Yet, I'm convinced that repeated courses of broad spectrum antibiotics are the main "villain." These antibiotics cause yeast overgrowth in your intestinal tract and vaginal yeast infections. And these infections, like a stream cascading down a mountain, set off disturbances that can make you feel "sick all over."

The Diagnosis of a Yeast-Related Disorder

How does your physician make a diagnosis? How does she find out **why** you're troubled by fatigue, headache, depression or other symptoms? She bases her conclusions on:

- **Your history or story.** This includes not only your main complaints, but also symptoms or events in your past that may be important.
- **Your physical examination.** This should include an overall look at your body, along with closer examination of your skin, eyes, heart, lungs and other parts of your body.
- **Laboratory examinations, other tests and X-rays.**

It's usually harder to diagnose a yeast-related problem than a fractured leg, pneumonia, or many other problems. And a physical examination and tests don't provide the information your physician needs to make a diagnosis.

Nevertheless, if you feel "sick all over," you should go to a physician for a checkup. You'll need to make sure your symptoms aren't caused by some other disorder.

If your physical examination and "routine" laboratory tests are normal, and your history suggests a yeast-related problem, *the diagnosis is made by noting your response to a simple, but comprehensive, treatment program. Such a program features a sugar-free special diet and prescription or nonprescription antiyeast medications.* (See also pages 200–205.)

MEDICAL HISTORY

* Repeated antibiotics during childhood
* Acne during teen years
* Recurrent urinary infections
* Frequent vaginitis
* Jock itch
* Sexual dysfunction
* Fatigue, depression, headache
* Visits to many different physicians
* Digestive symptoms
* Crave sweets
* Sick all over, etc., etc.

Candida Questionnaire and Score Sheet

If you'd like to know if your health problems are yeast-connected, take this comprehensive questionnaire.

Questions in Section A focus on your medical history—factors that promote the growth of *Candida albicans* and that frequently are found in people with yeast-related health problems.

In Section B you'll find a list of 23 symptoms that are often present in patients with yeast-related health problems. Section C consists of 33 other symptoms that are sometimes seen in people with yeast-related problems—yet they also may be found in people with other disorders.

Filling out and scoring this questionnaire should help you and your physician evaluate the possible role *Candida albicans* contributes to your health problems. Yet, it will not provide an automatic "yes" or "no" answer.

Section A: History

	Point Score
1. Have you taken tetracyclines or other antibiotics for acne for 1 month (or longer)?	35
2. Have you at any time in your life taken broad-spectrum antibiotics or other antibacterial medication for respiratory, urinary or other infections for two months or longer, or in shorter courses four or more times in a one-year period?	35
3. Have you taken a broad-spectrum antibiotic drug—even in a single dose?	6
4. Have you, at any time in your life, been bothered by persistent prostatitis, vaginitis or other problems affecting your reproductive organs?	25
5. Are you bothered by memory or concentration problems—do you sometimes feel spaced out?	20
6. Do you feel "sick all over" yet, in spite of visits to many different physicians, the causes haven't been found?	20

Candida Questionnaire and Score Sheet—*Cont'd.*

Section A: History—*Cont'd.*

	Point Score
7. Have you been pregnant . . .	
Two or more times?	5
One time?	3
8. Have you taken birth control pills . . .	
For more than two years?	15
For six months to two years?	8
9. Have you taken steroids orally, by injection or inhalation?	
For more than two weeks?	15
For two weeks or less?	6
10. Does exposure to perfumes, insecticides, fabric shop odors and other chemicals provoke . . .	
Moderate to severe symptoms?	20
Mild symptoms?	5
11. Does tobacco smoke *really* bother you?	10
12. Are your symptoms worse on damp, muggy days or in moldy places?	20
13. Have you had athlete's foot, ring worm, "jock itch" or other chronic fungous infections of the skin or nails? Have such infections been . . .	
Severe or persistent?	20
Mild to moderate?	10
14. Do you crave sugar?	10
TOTAL SCORE, Section A	

Section B: Major Symptoms

For each of your symptoms, enter the appropriate figure in the Point Score column:

If a symptom is **occasional or mild** . . . . 3 points
If a symptom is **frequent and/or moderately severe** . . . . . . 6 points
If a symptom is **severe and/or disabling** . . . . 9 points

Add total score and record it at the end of this section.

	Point Score
1. Fatigue or lethargy	
2. Feeling of being "drained"	
3. Depression or manic depression	
4. Numbness, burning or tingling	
5. Headache	
6. Muscle aches	
7. Muscle weakness or paralysis	
8. Pain and/or swelling in joints	
9. Abdominal pain	
10. Constipation and/or diarrhea	
11. Bloating, belching or intestinal gas	
12. Troublesome vaginal burning, itching or discharge	
13. Prostatitis	
14. Impotence	
15. Loss of sexual desire or feeling	
16. Endometriosis or infertility	
17. Cramps and/or other menstrual irregularities	
18. Premenstrual tension	
19. Attacks of anxiety or crying	
20. Cold hands or feet, low body temperature	
21. Hypothyroidism	
22. Shaking or irritable when hungry	
23. Cystitis or interstitial cystitis	
TOTAL SCORE, Section B	

Section C: Other Symptoms

For each of your symptoms, enter the appropriate figure in the Point Score column:

If a symptom is **occasional or mild** .... 1 point
If a symptom is **frequent and/or moderately severe** ...... 2 points
If a symptom is **severe and/or disabling** .... 3 points

Add total score and record it at the end of this section.

Candida Questionnaire and Score Sheet—*Cont'd.*

Section C: Other Symptoms—*Cont'd.*

	Point Score
1. Drowsiness, including inappropriate drowsiness	
2. Irritability	
3. Incoordination	
4. Frequent mood swings	
5. Insomnia	
6. Dizziness/loss of balance	
7. Pressure above ears . . . feeling of head swelling	
8. Sinus problems . . . tenderness of cheekbones or forehead	
9. Tendency to bruise easily	
10. Eczema, itching eyes	
11. Psoriasis	
12. Chronic hives (urticaria)	
13. Indigestion or heartburn	
14. Sensitivity to milk, wheat, corn or other common foods	
15. Mucus in stools	
16. Rectal itching	
17. Dry mouth or throat	
18. Mouth rashes, including "white" tongue	
19. Bad breath	
20. Foot, hair or body odor not relieved by washing	
21. Nasal congestion or postnasal drip	
22. Nasal itching	
23. Sore throat	

	Point Score
24. Laryngitis, loss of voice	
25. Cough or recurrent bronchitis	
26. Pain or tightness in chest	
27. Wheezing or shortness of breath	
28. Urinary frequency or urgency	
29. Burning on urination	
30. Spots in front of eyes or erratic vision	
31. Burning or tearing eyes	
32. Recurrent infections or fluid in ears	
33. Ear pain or deafness	
TOTAL SCORE, Section C	
Total Score, Section A	
Total Score, Section B	
GRAND TOTAL SCORE	

The Grand Total Score will help you and your physician decide if your health problems are yeast-connected. Scores in women will run higher, as seven items in the questionnaire apply exclusively to women, while only two apply exclusively to men.

Yeast-connected health problems are almost certainly present in women with scores **more than 180,** and in men with scores **more than 140.**

Yeast-connected health problems are probably present in women with scores **more than 120,** and in men with scores **more than 90.**

Yeast-connected health problems are possibly present in women with scores **more than 60,** and in men with scores **more than 40.**

With scores of less than 60 in women and 40 in men, yeasts are less apt to cause health problems.

New Support for Yeast-Connected Health Problems

Tens of thousands of people who wrote and called me during the mid and late 1990s said, *"I can't find a physician to help me and my own doctor is skeptical."*

I can understand this skepticism and here is one of the reasons. My original yeast book, the best-selling, *The Yeast Connection—*

A Medical Breakthrough, was first published in 1983 and hasn't been significantly revised since 1986. Since that time, only minor changes have been made.

However, a tremendous amount of new information became available in the late 1980s and 1990s which provides medical support for the relationship of *Candida albicans* to asthma, autism, endometriosis, interstitial cystitis, multiple sclerosis, psoriasis and other disorders. You'll find a brief summary of some of this information elsewhere in this book.

In the summer and fall of 1999, the International Health Foundation (IHF) published three booklets for people who write and call seeking information and help. (See pages 238–240.)

References

1. Iwata, K. and Uchida, K., "Cellular Immunity in Experimental Fungus Infections in Mice: The influence of infections in treatment with a candida toxin on spleen lymphoid cells," *Mykosen, Suppl.* 1, 72–81 (1978), Symposium Medical Mycology, Flims, January 1977.

2. Miles, M. R., Olsen, L. and Rogers, A., "Recurrent Vaginal Candidiasis: Importance of an Intestinal Reservoir," *JAMA,* 238:1836–1837, October 28, 1977.

3. Witkin, S. S., "Defective Immune Responses in Patients with Recurrent Candidiasis," *Infections in Medicine,* May/June 1985, pp. 129–131.

4. Brodsky, J. H., as quoted in the Foreword of *The Yeast Connection,* Third Edition, paperback, by W. G. Crook, M.D., Professional Books, Jackson, TN and Vintage Books, New York, 1986.

CHAPTER · TWO

People Who Feel Sick All Over

■ ■ ■

During the past 15 years I've received thousands of letters and phone calls from people with complaints that focused on a "disease" or a particular part of their body. Their complaints included fatigue, headache, depression, endometriosis, interstitial cystitis, multiple sclerosis and psoriasis. Yet, I've received an even greater number of calls and letters from people who, when asked about their medical history, said, "It's hard to know where to begin . . . I feel sick all over."

These letters came from men and women of all ages who were concerned about their own problems. Some letters came from parents, concerned about their children. The largest number of letters—some 85 percent—were from women, and the majority of them were between the ages of 25 and 50.

If you are such a person, it *is* important for you to seek out and find a physician who is not only competent, but also caring and compassionate. During the course of an examination or evaluation, he or she usually will be able to rule out many diseases and disorders. Yet, even a careful and comprehensive examination may fail to come up with a "diagnosis."

In discussing this situation, Martin H. Zwerling, M.D., Kenneth N. Owens, M.D., and Nancy H. Ruth, R.N., B.S., commented:

> Consider the following "incurable" patient who is being treated by several specialists. Her gynecologist is treating her recurrent vaginitis and irregular menstrual periods while an otolaryngologist is trying to control her external otitis and chronic rhinitis.

At the same time, an internist is unsuccessfully attempting to manage symptoms of bloating, indigestion and abdominal pain and a dermatologist is struggling with bizarre skin rashes, hives and psoriasis.

Lastly, psychiatrists have been unable to convince the patient that her nerves are the cause of her extreme irritability, inability to concentrate and depression. We've all been guilty of labeling such patients as "psychosomatic" and since there is "nothing physically wrong," conclude that we cannot cure them.

Incurable? Not if you think yeast. This patient and thousands like her are suffering from chronic candidiasis.

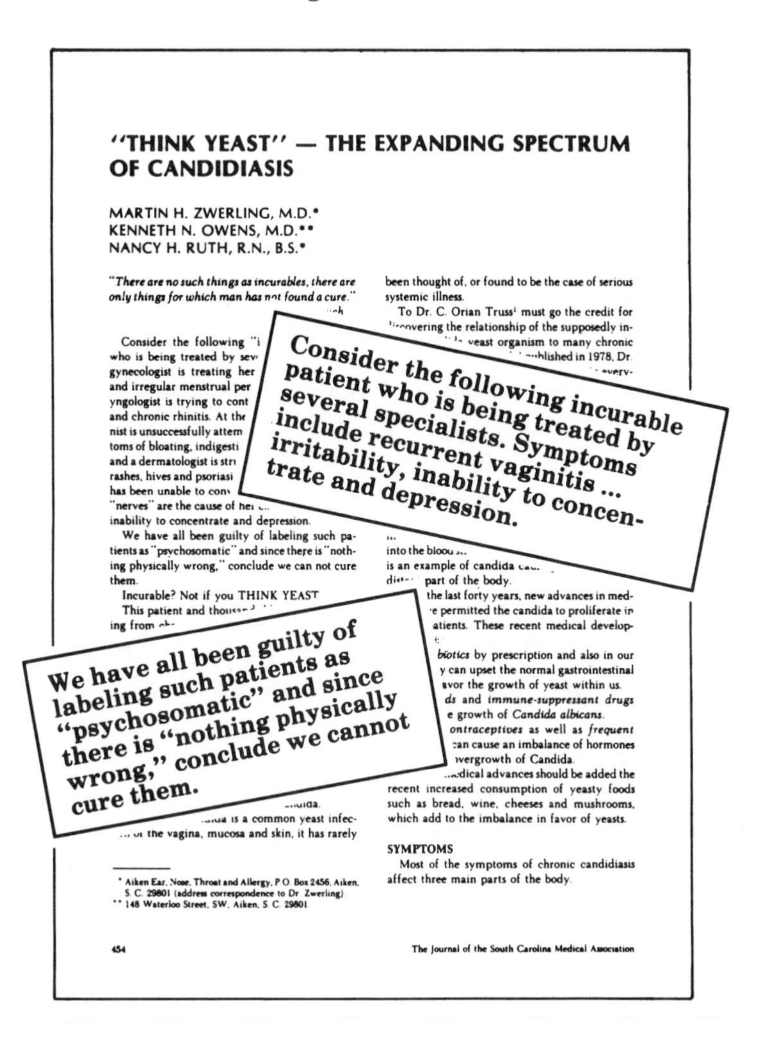

"THINK YEAST" — THE EXPANDING SPECTRUM OF CANDIDIASIS

MARTIN H. ZWERLING, M.D.*
KENNETH N. OWENS, M.D.**
NANCY H. RUTH, R.N., B.S.*

"There are no such things as incurables, there are only things for which man has not found a cure."

Consider the following "i
who is being treated by sev
gynecologist is treating her
and irregular menstrual per
yngologist is trying to cont
and chronic rhinitis. At the
nist is unsuccessfully attem
toms of bloating, indigesti
and a dermatologist is stri
rashes, hives and psoriasi
has been unable to conv
"nerves" are the cause of her
inability to concentrate and depression.

We have all been guilty of labeling such patients as "psychosomatic" and since there is "nothing physically wrong," conclude we can not cure them.

Incurable? Not if you THINK YEAST

This patient and thous
ing from

is a common yeast infec-
of the vagina, mucosa and skin, it has rarely

been thought of, or found to be the case of serious systemic illness.

To Dr. C. Orian Truss[1] must go the credit for
covering the relationship of the supposedly in-
yeast organism to many chronic
published in 1978, Dr.

into the blood
is an example of candida
part of the body.
the last forty years, new advances in med-
e permitted the candida to proliferate in
atients. These recent medical develop-

biotics by prescription and also in our
y can upset the normal gastrointestinal
avor the growth of yeast within us.
ds and *immune-suppressant drugs*
e growth of *Candida albicans.*
ontraceptives as well as *frequent*
can cause an imbalance of hormones
overgrowth of Candida.
dical advances should be added the recent increased consumption of yeasty foods such as bread, wine, cheeses and mushrooms, which add to the imbalance in favor of yeasts.

SYMPTOMS

Most of the symptoms of chronic candidiasis affect three main parts of the body.

Consider the following incurable patient who is being treated by several specialists. Symptoms include recurrent vaginitis ... irritability, inability to concentrate and depression.

We have all been guilty of labeling such patients as "psychosomatic" and since there is "nothing physically wrong," conclude we cannot cure them.

* Aiken Ear, Nose, Throat and Allergy, P.O. Box 2456, Aiken, S. C. 29801 (address correspondence to Dr. Zwerling).
** 148 Waterloo Street, SW, Aiken, S. C. 29801

454 The Journal of the South Carolina Medical Association

Typical Stories

Deborah

This 38-year-old woman called my office for an appointment in October 1984. Here are excerpts from a six-page letter she sent me before her visit.

> For over 15 years I've been bothered by recurrent urinary tract infections and persistent vaginal yeast infections. For at least 10 years I've had to make myself get out of bed every morning. I couldn't hear an alarm clock so my husband would literally have to drag me out. Many days I would feel tired and dazed, like a rag doll or a wet washrag.
>
> Over the years I've been to all sorts of doctors because of my many complaints, including aching in my neck, joints and legs, noises in my ears, constipation, bloating, poor memory and a feeling of being "spaced out." As a result of all of these symptoms I've had absolutely no interest in sex. When the doctors could not find a cause for these symptoms, several of them said, "See a psychiatrist."
>
> I've been hospitalized several times, but the tests never showed anything. Chemicals of all sorts bothered me, including several brands of perfume. My symptoms also get worse when I eat sweetened foods or drink milk.

I put Deborah on a sugar-free, milk-free special diet, nystatin and nutritional supplements. She also got rid of the odorous chemicals in her home. Two weeks after beginning treatment, Deborah reported:

> I'm much better. Less vaginal itching and burning, and less frequent urination. My energy level has improved significantly. I no longer feel bloated.

Six months after she started treatment she reported that she was symptom free and well—except when she cheated on her diet. During the past few years Deborah has continued to do well, and in March 1993, she wrote that she was happy and healthy. She rarely experienced symptoms except when she was exposed to chemicals or consumed too many sweets.

Mike

I took tetracycline for two years because of acne. In my early '20s I developed headache, fatigue, abdominal pain and bloating. I just didn't feel good.

Checkups by my family doctor and by a gastroenterologist who did an endoscopic examination showed no abnormalities. So I began to feel like a hypochondriac. I felt that a 28-year-old guy with a darn good job and a beautiful girlfriend should feel a lot better than I did. But I didn't know where to turn.

Then, in browsing through the book section of a health food store, I came across your book, *The Yeast Connection,* and I was very impressed. I changed my diet, eliminating sugar and most yeast products. I also started taking Mycopryl—a caprylic acid product—and I've noticed a marked improvement in my general health.

Thanks so much. I just wanted you to know that your book has really changed my life.

David's Wife

My husband had been ill for 13 years and had been to doctor after doctor. For 10 years they treated him for arthritis with endless new drugs and steroids. Then he started developing sensitivity to tobacco smoke, perfumes and various chemical odors. He was also troubled by headaches and stomach problems. At times we thought he was losing his mind.

Doctors could not find one thing wrong; yet, David continued to grow weaker and more discouraged. Then a friend told us about your book, *The Yeast Connection.* We read it from cover to cover and found a knowledgeable, caring doctor. He put David on a sugar-free special diet and he gave him a prescription for nystatin.

He also outlined a program of nutritional supplements, including vitamins, minerals and flaxseed oil. David improved even more when using elimination diets. We found that David is sensitive to milk and wheat. Thank you for changing our lives.

Paige Grant

I enjoyed fairly good health except for occasional vaginal infections and headaches, which had plagued me since I was a child. Then in my early 30s, I developed a number of urinary tract infections, but I was able to control them with antibiotics. In spite of these problems, I was a full-time model and commercial actress and I never missed a day's work from illness.

Then in my mid-30's I began to subtly slide into various problems. By 1990 I had very little energy, felt generally awful, slept a lot and was seeking medical help frequently. I started to feel like I was crazy because the doctors couldn't find out exactly what was wrong with me. They would usually give me antibiotics and I would get better, then I would get worse.

Since 1984 I estimate that I saw about 50 different doctors for various ailments, including a psychotherapist, because I was beginning to think it might be "all in my head." My husband and I estimate my cost for this period at about $100,000.

Finally I saw Dr. Alan McDaniel, a New Albany, Indiana, ear, nose and throat specialist and an allergist. I think he saved my life. On a comprehensive program, which included anticandida therapy and treatment for my allergies, I began to feel like a new person. Later I moved to Chicago and continued under the good care of Dr. George Shambaugh.

I've been involved in many personal and community projects, which I never would have been able to do until now. I no longer have the irritability and mood swings. I only have headaches now when I eat the wrong food.

I've had such a dramatic recovery that I would like to see other women reached by your work. Please let me know how I can help.

Louise

I'm a 66-year-old retired school teacher and I enjoyed good health until I was in a car accident two years ago. Although I suffered no serious injuries, I was pretty well banged up. My treatment included antibiotics of several sorts, which I took for many weeks. *Although my doctor discharged me and said, "You're fine," I didn't feel good.* I was tired, had daily headaches and my digestive system was all messed up. I had bloating, constipation and occasional diarrhea.

Then I developed a urinary tract infection and was given more antibiotics. Although the urinary symptoms improved, I developed a vaginal yeast infection for the first time in 30 years. My doctor recommended vaginal suppositories, which helped, but then the yeast infection came back, and I still felt tired and out of sorts. Then, a friend told me about *The Yeast Connection and the Woman*. I started the sugar-free diet and persuaded my doctor to prescribe nystatin and Diflucan. I'm already a lot better. Thanks for your help.

Jeri Coffey, DDS

During my 20s I craved sweets, had constant sore throats and didn't feel well. I experienced many illnesses and I had repeated headaches. In my efforts to find help I saw five different allergists. Nothing they did was effective. Then I was referred to Dr. Shambaugh, and two weeks after starting on a comprehensive program, which included nystatin and a sugar-free diet, I began to improve.

I also have two boys, ages 11 and 7, who in the past have experienced allergies of various sorts. Now, on a good diet and environmental control measures, they're healthy.

I feel so good now that I'm sorry I didn't feel this way when I was in my 20s. At one point Dr. Shambaugh said to me, "I hope that after you follow my treatment program, you'll feel as good as I do!"

Eleanor

I'm very sick and need help. The doctors around here refuse to believe anything about yeast-related illness. I've read books and have been treating myself with some things from a health food store. I get better, then I get bad again. I feel like I'm slowly being poisoned.

I've been in many doctors' offices and hospitals and they don't find anything wrong. Yet, I don't get well and all I've been given are more nerve pills and antidepressants, which make me sicker.

I have so many doctor and hospital bills piled up that if I could find a doctor to help me I really don't know how I would pay him. Very slowly I guess. If there's anyone in the country who can help me I'll travel there. I'm sick, I'm desperate, I'm broke and I'm scared—real scared. If there's research going on anywhere, I would volunteer to be a part of it.

Here's a list of my symptoms I've had for the last 10 years, and the last six months these symptoms have been worse:

vaginal infections	pain in jaw	PMS
depression	dizziness	thrush
muscle and joint pain	always tired	headache
pain down back	tightness in chest	nausea
sinusitis	shortness of breath	bloating
constipation	no feeling in vagina	colitis

> If you could help me in any way I would be grateful. I want to stop this before it goes into something more serious. I have a lot to live for and I just want to get well.

Barbara

This three-year-old girl was seen by Bruce Semon, M.D., Ph.D. with the complaint of speech delay. Dr. Semon, a child psychiatrist and nutritionist, has been treating children with autism* using anti-yeast therapy for over six years. Here is a description of Barbara's symptoms and treatment as outlined in the 1998 book, *Biological Treatment for Autism and PDD,* by William Shaw, Ph.D.

> Barbara had seemed normal until about 18 months when she had a severe vaginal yeast infection which continued for six months. At that time, her speech stopped developing. Her parents reported she could only say a few words. She was unhappy and screamed occasionally and ran around.
>
> She was started on diet and nystatin. She came back three weeks later. Her parents reported that she was attending school and could sit still for longer periods of time and was having fewer temper tantrums. In my office, she smiled, played with a toy and was not hyperactive.

Dr. Semon, the author of a 25-page chapter in Dr. Shaw's book, describes his experiences in treating a number of autistic children. Here are excerpts from the two concluding paragraphs of his chapter.

> I would advise any parent of an autistic child to treat their child with diet and nystatin. . . . Following the diet and using nystatin for 2–4 weeks will tell you if the treatment is beneficial for your child with no adverse effects or risks to your child.

*In 1999, Dr. Semon and his wife published a cookbook entitled *Feast Without Yeast—4 Stages to Better Health.*

CHAPTER · THREE

Why Women Are Affected More Often Than Men

Women develop yeast-related problems more often than men, and premenopausal women appear to be especially susceptible. There appear to be several reasons for this:

Differences in Anatomy

Women are more likely than men to develop urinary and genital yeast infections because of their anatomy. For example, a woman's urethra (the tube leading from the urinary bladder to the outside) is short, making it easier for bacteria to enter the woman's bladder and set up an infection.

Urinary tract infections are especially apt to occur in women after frequent or prolonged sexual intercourse. These infections are usually treated with antibiotic drugs. The candida yeasts, which live normally in the intestinal tract, multiply when a woman takes antibiotics.

The proximity of the anal opening to the vulva and vagina increases a woman's chances of developing a genital infection. Since yeasts thrive on the warm, dark, interior membranes of the body, the vagina furnishes a hospitable home.

The Pill

In his classic book, *The Missing Diagnosis,* Dr. C. Orian Truss listed the various factors that cause candida-related health problems and their prevention. Included were antibiotics, dietary factors, immunosuppressive drugs and birth control pills. Here is an excerpt:

> ... Approximately 35 percent of women using "the pill" were having severe chronic yeast vaginitis . . . Alternate methods of contraception . . . should replace the use of these hormones in women susceptible to their impact on yeast growth.[1]

For many years, Ellen Grant, a British obstetrician/gynecologist, has expressed concern about possible adverse effects of the pill. In her book, *The Bitter Pill*, she said, "Candidiasis is at least doubled among pill users . . ."[2]

Hormonal Changes

Yeast colonization is encouraged by hormonal changes associated with the normal menstrual cycle during pregnancy. Quoting Dr. Truss again:

> Estrogen is produced throughout the monthly cycle. Progesterone is produced in very small quantities prior to ovulation, but thereafter in large quantity until the onset of the next period. This high level of progesterone persists throughout pregnancy when conception occurs . . . By unknown mechanisms, progesterone greatly aggravates yeast growth in women . . .[3]

Premenopausal Women Go to Physicians More Often Than Men

A young woman is more likely than a young man to develop a personal relationship with a physician because she sees her physician more often—for routine checkups and Pap smears, when she gets pregnant or when she has a vaginal yeast infection.

Accordingly, when she develops a fever, cough or cold, she's more likely to contact her physician to ask for relief. He may prescribe an antibiotic, which promotes the growth of yeasts.

Prolonged Antibiotics for Teenagers with Acne

Teenagers, especially girls, are concerned about their complexions. So they're more apt to consult a physician and to be put on long-term antibiotics. Although these drugs seem to help some youngsters with acne, they wipe out normal bacteria in the intestinal tract. As a result, yeasts multiply and a "cascade" of other health problems often develop.

During the past 15 years since my first yeast book was published in 1983, I've received tens of thousands of letters and phone calls on the IHF hotline (901-660-7090) which I answer most Tuesdays between 1:15 and 2:15 P.M. Central Standard Time. The majority of these calls come from women between the ages of 25 and 50 with complaints of being "sick all over." In responding to these calls, here's one of the first questions I ask:

*Did you, during your teen years, take long-term tetracycline for acne? And many of these people, including men, answer, "Yes."**

References

1. Truss, C. O., *The Missing Diagnosis,* P.O. Box 26508, Birmingham, AL 35226, 1993; p. 30.

2. Grant, E., *The Bitter Pill,* Corgi, Elm Tree/Hamish Hamilton Edition, London, 1986; p. 174.

3. Truss, C. O., *The Missing Diagnosis,* p. 30.

*See also pages 238–239.

CHAPTER · FOUR

Common Problems That Affect Adults of Both Sexes

Fatigue

I developed an interest in chronic fatigue, headache and other symptoms almost 40 years ago. At that time, an alert mother convinced me (against my will!) that her 12-year-old son's fatigue, weakness and other symptoms vanished when he stopped drinking milk.

A short time later, I read several articles in the medical literature that described food-related fatigue and other symptoms.[1,2,3] After reading these articles, I began putting a number of my tired, irritable patients on elimination or challenge diets. So many of them improved that I kept careful records of my observations and published them in a major pediatric journal in 1961.[4]

During the remainder of the 60s and on into the 70s, I helped hundreds of tired, irritable children, who also complained of headaches, muscle aches and other symptoms, by tracking down and eliminating foods that triggered their symptoms. I published more of my observations in the peer-reviewed literature.[5,6]

During the 1970s, I began to see more adult patients, especially women, who complained of fatigue and other symptoms that made them feel "sick all over." Although some of these patients improved when they changed their diets and cleaned up the pollutants in their homes, many remained ill.

I saw one such patient, 35-year-old Linda, a number of times in 1976. Although she improved on the treatment program I recommended, she continued to experience many symptoms. Linda moved away and I lost contact with her until the fall of 1979. At that time, she told me that a special diet and antiyeast medication had helped her overcome her chronic fatigue and other symptoms. She gave me a copy of an article by Dr. Truss published in a little known Canadian medical journal.[7]

Soon afterward, I began to treat and help other patients, who complained of fatigue, headache, depression, muscle aches, poor memory and other symptoms, with a sugar-free special diet and the antiyeast medication, nystatin.

In the 1980s, I saw hundreds of adults with yeast-related problems. And in preparing for a presentation at a medical conference, I reviewed the histories of 100 consecutive new patients. In collecting statistical data, I found that fatigue was the chief complaint in most of these patients. Headache was second and depression third.

Chronic Fatigue Syndrome (CFS)

During the mid-1980s, reports about chronic fatigue syndrome (CFS) began appearing in medical journals and in the press. Professionals and nonprofessionals sought answers to these questions:

- How does chronic fatigue syndrome start?
- Is there more than one cause?
- Is it contagious?
- What makes it persist?
- What is the best treatment?

Many physicians, including researchers, emphasized the role of one or more viruses in causing this disorder. For a time, some considered the Epstein-Barr virus the culprit. Yet, laboratory and other studies showed that this virus was not the causative agent. Although CFS developed suddenly in many people after a viral infection, it seemed to come on more gradually in others.

As I read the reports in the medical and lay literature about CFS, it seemed to me that the symptoms in many people with this illness were similar to those experienced by my patients with yeast-related problems.

During the late 1980s and into the 90s, physicians and other professionals began to study this disorder. They expressed different opinions about the causes. Some observers blamed it on psychological causes; others blamed it on food and chemical sensitivities. Yet, most researchers interested in pursuing this felt that viral infections were the major causative agent—although a single culprit was never identified. And because people with CFS were found to have measurable disturbances in their immune systems, the term Chronic Fatigue and Immune Dysfunction Syndrome (CFIDS) was suggested as a more appropriate term.*

Observations of Others

At an April 1989 CFS conference sponsored by the University of California, San Francisco, and at a November 1990 CFS/CFIDS conference in Charlotte, North Carolina, Dr. Carol Jessop discussed her findings in working with more than 1,100 patients with CFS during the 1980s. Her treatment program featured the use of a restricted diet (no alcohol, sugar, fruit or fruit juice) and the antifungal medications Nizoral and Diflucan. In summarizing her experiences, she said:

> I don't believe we're going to find one single virus that causes this illness. I feel that, in addition to viruses, genetic predisposition and environmental factors, such as antibiotics, birth control pills, toxins in the environment and infections, all have to be considered.
>
> Eighty-four percent of my CFS patients have recovered to a level where they can remain working 30 to 40 hours a week and 30% have fully recovered. Yet, 44% of the patients experience some recurrence of their symptoms with premenstrual stress, surgery or other infections.

A number of other physicians including Charles Lapp of Raleigh, North Carolina, Jacob Teitelbaum of Annapolis, Maryland, Michael Goldberg, Tarzana, California, Philip Nelson, Sarasota,

*To obtain more information, call or write the CFIDS Association of America, P.O. Box 220398, Charlotte, NC 28222, 1-800-44-CFIDS (1-800-442-3437).

Florida have observed that their patients with CFS, CFIDS can be helped by a comprehensive program which includes antifungal medication.

In discussing his experiences, Dr. Goldberg said that he helped hundreds of his CFS/CFIDS patients using Diflucan and/or other antifungal medications. Patients most apt to respond favorably had local yeast infections, digestive symptoms and aching muscles.[8]

Dr. Nelson, the medical advisor of a Florida CFS/CFIDS support group, commented:

> *I must admit for years I was a complete skeptic. Several things have changed my mind.* I've had the opportunity of working with chronic fatigue syndrome patients over the past two years, and seeing their response to both diet and antifungals has been impressive.[9]

An Exciting Story

Virginia is a teen-aged patient who came to me in 1992. Because of acne, Virginia had been put on tetracycline. A short time later she developed extreme fatigue, headaches, depression and generalized body aching. Unable to go to school, she was given a homebound teacher.

On a treatment program that included a sugar-free, dairy-free, special diet and antifungal medication, Virginia improved steadily. She experienced a setback after taking antibiotics for a urinary tract infection, but following dietary changes and anti-fungal medications, her symptoms subsided. Virginia is now a healthy, happy young woman and doing well in college.

Fibromyalgia Syndrome (FMS)

Individuals who suffer from Fibromyalgia Syndrome (FMS) are troubled by neck and shoulder pain, morning stiffness, sleep disturbances and fatigue. They also have many tender points in various parts of their body. Recent estimates suggest that three to six million people in this country are troubled by FMS. Seventy to 90 percent are women; the mean age for diagnosis has varied from 34 to 55.[10]

At several conferences in the early and mid-90s, professionals who discussed CFS/CFIDS and FMS expressed varying points of view and recommended different treatments. The general consensus, however, seemed to be that these different syndromes were closely related—if not the same disorder.*

My Comments

Almost ten years before chronic fatigue syndrome began to be talked about in medical journals and the media, C. Orian Truss, an Alabama internist, published an article describing mental and nervous system manifestations caused by superficial yeast infections. The symptoms in these patients included fatigue, lethargy, memory loss, poor concentration, headaches, muscle pains and other symptoms.

In an article published in the Fall 1992 issue of *the Journal of Advancement in Medicine,* Truss reported on his continuing experiences in studying women with superficial candida infection of the vagina and gastrointestinal tract accompanied by generalized symptoms. He again emphasized the relationship of yeast to chronic fatigue syndrome.[11]

Based on my own experiences in practice, and those of other physicians, including Drs. Jessop, Goldberg and Nelson, I feel that CFS/CFIDS and FMS are often yeast related. People with these disorders seem to develop them because their immune systems are disturbed. When their immune systems are weakened, viruses are activated, yeasts multiply, food and chemical allergies become activated, and nutritional deficiencies develop.

Each person is unique. Some individuals will be bothered more by fatigue, headache and depression; others will experience more muscle aching. Still others will be bothered especially by cognitive dysfunction and/or other symptoms.

I'm not claiming that the common yeast, *Candida albicans,* is **the** cause of CFS/CFIDS/FMS and other disorders that affect many people. Yet, increasing evidence shows that a sugar-free special diet and antifungal medications may help many people with these chronic health disorders get well.

---

*You'll find more information in the newsletter, *Fibromyalgia Network,* P.O. Box 31750, Tucson, AZ 85751–1750, 1-800-853-2929. Published quarterly, $19.

Headaches

Headaches rank ninth among the causes of visits to physicians and are a major source of both time lost from work and of medical diagnostic procedures, according to Jerome B. Posner,[12] professor of neurology at Cornell University Medical College.

In my pediatric and allergy practice during the past 35 years, many of my patients—both children and adults—gained relief from their headaches when they avoided one of their common foods. I published my observations for the first time in a report of 50 patients in 1961.

In a subsequent article published in 1975, I again told of my experiences in helping youngsters with persistent headaches with a simple treatment program that featured dietary changes.

Many physicians during the past 50 years have described the relationship of diet to headaches of various types. Among the reports, which especially interested me, were those of British observers, Ellen Grant[13] and Joseph Egger.[14]

Headaches and the Yeast Connection

Several years ago, when I analyzed the histories of 100 women with yeast-related problems, fatigue, headache and depression were at the top of the list. And during these past 10 years, I've received countless letters from people with the major complaint—*headache*.

In the summer of 1992, as I was gathering material for a presentation in the Chicago area, I called my good friend, Dr. George E. Shambaugh, Jr., and asked him if he could give me the names of patients he had helped by using antifungal medications and a special diet. One story that was especially dramatic came from Kathleen Pawalczyk, the mother of one of Dr. Shambaugh's patients.

Kathleen Pawalczyk's Story

My daughter, Rachel, started having headaches and stomach aches after a bout with the flu in December 1986. She also had the urge to urinate all the time and complained of being tired. She developed asthma and would get bronchitis every October.

The doctors treated her with numerous antibiotics and other medications. She gradually became worse and worse.

We took her to numerous specialists . . . No doctor could find a medical reason for all her complaints. In the 1989–90 school year, she missed 97 days of school. *She would lie on the couch all day and dunk her head in a bowl of ice water for relief from her headache.*

About that time, Kathleen learned that yeast overgrowth could make people sick all over. A friend told her about Dr. Shambaugh, who put Rachel on a simple comprehensive treatment program. It featured a sugar-free, special diet and antiyeast medications. When I called Rachel in October 1995, she said, "I'm healthy, attend school regularly, and my headaches and other symptoms are a thing of the past."

My Comments

How and why are headaches yeast connected? Several mechanisms are probable, but here's the one that I think is most important: Repeated courses of antibiotics knock out many of the friendly bacteria in the intestinal tract. Yeasts and unfriendly bacteria multiply and raise large families. These changes lead to a "leaky gut" and the absorbtion of food allergens and toxins.

Depression* and Manic Depression

Millions of Americans suffer from depression. Many of them who went years without relief are receiving help from Prozac, Zoloft and other medications. Yet, in spite of their benefits, these medications, like many others, may cause adverse reactions.

Because of these possibilities, many therapists say that these drugs *should not be given to people without looking for the underlying physical and psychological causes of the person's depression.*

Like fatigue, headache, PMS and other chronic complaints, depression may develop from many different causes. Moreover, such causes are often multiple. In my experience and that of many other practicing physicians *Candida albicans* ranks high on the list . . . especially in women between the ages of 25 and 45.

In his book, *The Missing Diagnosis,* C. Orian Truss, M.D., noted that depression *in many of his patients responded to anticandida therapy . . . often dramatically.* Yet, he cautions:

> Depression is a serious and potentially dangerous condition and one deserving care by a competent psychiatrist; self-diagnosis and treatment should never be attempted. *It is perfectly reasonable to look for some correctable cause of depression, but even when found, its treatment should not immediately and abruptly replace the psychiatric program.*
>
> Drugs prescribed by the psychiatrist should be gradually withdrawn, preferably under his supervision. Their use should not be discontinued suddenly or prematurely.[15]

Manic Depression

During the past few years, I've received a number of letters from people with manic depression who responded to antifungal therapy. Here are excerpts from a New Zealand woman's story, which I found to be especially dramatic:

---

*See page 254 for a 1996 report on the use of antifungal medication in women with depression.

> Imagine the whole world spray painted gray or being in a small windowless cell or in a tunnel. I felt that I had 50-pound weights on each foot and all my favorite things suddenly became meaningless and sterile.
>
> If someone had given me two round-trip air tickets to London and Paris and $15,000 spending money, I would have been completely unmoved. Nothing could trigger a flicker of interest or enthusiasm . . .

Then, after a week or two of this hell, she would swing into a manic world of exhaustion and even delirium.

> I was king of the castle, drunk with joy, bursting with crazy schemes, talking nonstop, spending money like water, issuing dinner and party invitations, smashing up the car. You'd have to have seen it to believe it . . .

This woman struggled with her depression for eight long years. She was put into a psychiatric hospital and received out-patient therapy for many months.

Then a general practitioner prescribed a diet, nystatin and nutritional supplements. On this treatment program, the bouts of mania stopped immediately and the depressions became brief and much less severe; within three months they disappeared.

My Comments

Candida albicans is not *the* cause of depression. Yet, if you suffer from depression and/or any other disabling disorder and have a history of . . .

- Repeated or prolonged courses of antibiotic drugs
- Persistent digestive symptoms
- And/or recurrent vaginal yeast infections, prostate infections, jock itch or nail fungus . . .

. . . a comprehensive treatment program that features oral antifungal medications and a special diet may provide you with sometimes dramatic relief from recurring depressions.

Multiple Chemical Sensitivity Syndrome (MCSS)

During the 1940s, Theron Randolph, a Chicago internist/allergist, found that many of his allergy patients improved when they avoided several of their favorite foods, including corn, wheat and dairy products. Yet, there were other patients who continued to experience problems.

Then in the 1950s, he noted that the symptoms in many of his patients were triggered or aggravated by chemical exposures.[16]

Gradually, during the past 30 years, a handful of other physicians began to recognize and treat patients with chemical sensitivities. Yet, most members of the medical "establishment" weren't interested.

I first learned about Randolph's observations many years ago. Then, during a visit to Chicago, he invited me to accompany him while he visited his patients in the hospital. So, as was my usual custom, I bathed, shaved and doused my face with cologne. I met Dr. Randolph in the lobby of the hospital to make "rounds."

The first patient was a woman who had been hospitalized because of asthma. She was convalescing and was sitting comfortably in a chair. Dr. Randolph introduced me, and I moved over and shook her hand. Within a minute or less, she began to cough; then, she began to wheeze because my cologne had triggered her symptoms. Dr. Randolph rushed me out of her room and sent me back to my hotel to take a bath and put on fresh scent-free clothing.

My Observations in Practice

During the ensuing years as I learned more about Dr. Randolph and his observations, I found that many of my patients improved after they stopped using colognes, household cleaners and other volatile chemicals in their homes.

Then in the early 1980s, after learning of Dr. Truss' observations, I found out that I was able to help many of my chemically sensitive patients by adding antifungal medications to their treatment program.

I reported my observations to my colleagues at a staff conference at Jackson-Madison County General Hospital and I published them in my state medical journal. In my article, I pointed out that many people with food and chemical sensitivities improved on antiyeast medications and a special diet. I also cited Truss' comments and a 1978 paper in which he said,

> We can conclude that one of the subgroup of patients with chronic candidiasis consists of those with severe intolerance to virtually all chemicals . . . One third of these patients have been found to have low T-cells . . . It has become increasingly apparent that these chemical sensitivities are disappearing and the T-cells are returning to normal (following treatment with nystatin) . . . indicating that the low T-cell counts were caused by *Candida albicans*.[17]

Other Stories

During the past 15 years, I've received countless letters, phone calls and reports from people of both sexes whose health problems developed after exposure to chemicals in their home or workplace.

The next story told by Diane Thomas is typical. It illustrates the combined effects of chemicals and yeast in causing debilitating health problems.

Diane's Story

Except for bad allergies to grass and ragweed as a young child, I was generally healthy. Yet, I was bothered from time to time with urinary problems. My mother, who had been a nurse, was sure they were allergy related. Every time my bladder would get upset, she would take me off citrus fruits, colas and tea, then I'd get better.

Although dietary changes would sometimes help, on other occasions I'd be put on antibiotics. Then, in my mid-30s, things began to get worse and when I was 36, I was told I had interstitial cystitis. I was also bothered by recurring vaginal yeast infections.

Then about that time I bought an old house and began restoring it. The restoration involved knocking out a lot of plaster and using odorous chemicals. And one of my neighbors a couple of times a month would spray malathion to kill the weeds along his fence, which was about ten feet from one of my windows.

Although I didn't realize the connection between the chemical exposures and my health problems, I developed many new symptoms, including fatigue and PMS. My uterus felt like someone had taken it out, jumped up and down on it real hard and then put it back. It was very, very tender.

In her continuing discussion, Diane told me that she had repeated bladder problems and took more antibiotics. She also had a hysterectomy, after which her doctor prescribed Keflex. Then she developed panic attacks. She said:

I emerged sensitive to everything. Then, about 13 years ago, I learned about Dr. Truss. He put me on his program, including a special diet and nystatin, and I began to improve. I also began taking acidophilus and many nutritional supplements.

Although it's been an uphill battle, I'm now a reasonably healthy, productive person. I have a good job, a good marriage and when I compare my health to what it was ten years ago, I realize that I'm really well off.

My Comments

In reviewing Diane's story, I can see that her health problems were obviously related to chemicals. And if we compare them to the "overloaded camel," her chemical bundle of straw was a heavy one. Yet, because she took many courses of antibiotics, yeast overgrowth was also an important factor.

References

1. Rowe, A., "Allergic Toxemia and Migraine Due to Food Allergy," *California and Western Medicine,* 1930; 33:785.

2. Randolph, T., "Allergy as a Cause of Fatigue, Irritability and Behavior Problems in Children," *J. of Ped.,* 1947; 31:560.

3. Speer, F., "The Allergic-Tension-Fatigue Syndrome," *Ped. Clin. of N. Amer.,* 1954; 1:1029.

4. Crook, W. G., "Systemic Manifestations Due to Allergy," *Pediatrics,* 1961; 27:790.

5. Crook, W. G., "The Allergic-Tension-Fatigue Syndrome," *Pediatric Annals,* October 1974.

6. Crook, W. G., "Food Allergy—The Great Masquerader," *Ped. Clin. of N. Amer.,* 1975; 22:227.

7. Truss, C. O., "Tissue Injury Induced by *C. Albicans:* Mental and Neurologic Manifestations," *J. of Ortho. Psych.,* 1978; 7:17–37.

8. Goldberg, M., Personal communication, 1994.

9. Nelson, P., Personal communication, 1994.

10. Goldenberg, D. L., "Fibromyalgia Syndrome—An emerging but controversial condition," *JAMA,* 1987; 257:2782–2787.

11. Truss, C. O., Truss, C. V., Cutler, R. B., "Generalized Symptoms in Women with Chronic Yeast Vaginitis: Treatment with nystatin, diet and immunotherapy versus nystatin alone," *J. of Advan. in Med.,* 1992; 5:139–175.

12. Posner, J. D., M.D., *Cecil's Textbook on Medicine,* W. B. Saunders Co., Philadelphia, PA, 1988; p. 2129.

13. Grant, E. C., "Food Allergies and Migraine," *The Lancet,* 1979; 1:986–988.

14. Egger, et al, "Is Migraine Food Allergy? A double-blind control trial of oligoantigenic diet treatment," *The Lancet,* 1983; 2:865–868.

15. Truss, C. O., *The Missing Diagnosis,* P.O. Box 26508, Birmingham, AL 35226, 1986; pp. 73, 75–76.

16. Randolph, T. G., *Human Ecology and Susceptibility to the Chemical Environment,* Charles C. Thomas, Springfield, IL, 1962.

17. Crook, W. G., "The Coming Revolution in Medicine," *J. of Tenn. Med. Assoc.,* 1983; 76:145–149.

CHAPTER · FIVE

Other Disorders That Affect Men and Women

■ ■ ■

Asthma

During the past several years, I've received a number of reports from people with chronic asthma who improved significantly—even dramatically—when they took antifungal medications. And in digging through the medical literature, I found several reports that described the relationship of *Candida albicans* to asthma. Interestingly enough, although two of these reports were published in the 1960s in major allergy journals,[1, 2] most allergists expressed little interest in this relationship.

Then, in a 1987 article in the *Annals of Allergy,* P. Gumowski and his colleagues noted that long-term corticosteroid therapy and antibiotics could play a part in causing asthma. They stated:

> Besides specific clinical pictures due to the pathogenicity of *Candida albicans* in cases of immune deficiencies, this yeast is known as an important causative allergen in bronchial asthma, chronic rhinitis, chronic urticaria, and food intolerance.[3]

Then, in 1994, other researchers noted that antifungal medications helped some patients with asthma.

Researchers at the University of Virginia* noted that some of their patients with fungal infections of the feet improved when given Diflucan. Following this course of therapy, eight of the 10 patients were able to reduce their dose of steroids.[4]

---

*These researchers published further observations in an article in the *Journal of Allergy and Clinical Immunology* (1999; 104:541-6).

A second abstract, published in the same issue of *The Journal of Allergy and Clinical Immunology,* came from Belgian researchers who also described the response of some of their asthmatic patients to the systemic antifungal drug, Nizoral.

Ten corticosteroid-dependent asthmatic patients, who showed no evidence of a fungal infection, were entered into a double-blind, placebo-controlled study.

These observers found that four out of five of the treated patients improved after two weeks, while four out of five of the placebo group did not improve. No side effects of the drug were observed. And they stated:

> We conclude that ketoconazole might be beneficial in some asthmatic patients. Further studies are needed to investigate the mechanism of action and a possible steroid sparing effect of ketoconazole in asthma.[5]

My Comments

Asthma, like many other chronic disorders, develops from multiple causes. Yet, I feel that any person with chronic asthma who has received repeated antibiotic drugs and/or corticosteroids, should be given a trial of oral antifungal medications and a sugar-free special diet.

Psoriasis

At a June 1982 medical conference in Dallas, Sidney M. Baker, M.D., who at that time was a member of the clinical faculty of Yale University School of Medicine, described the favorable response of several of his patients with psoriasis to oral nystatin.[6]

Word of Baker's observations spread to E. W. Rosenberg, M.D., professor and chairman, Division of Dermatology, University of Tennessee Center for Health Sciences in Memphis. In a letter to the editor published in the *New England Journal of Medicine,* Rosenberg said:

> We have been aware . . . of improvement of both psoriasis and inflammatory bowel disease in patients treated with oral nystatin, an agent that was expected to work only on yeast in the gut lumen. We've now confirmed that observation in several

of our patients with psoriasis. We suspect, therefore, that gut yeast may have a role in some instances of psoriasis.[7]

Subsequently, these observers and their colleagues published further observations on the role of yeast to psoriasis and the response of many of their patients to antifungal medications. And they've found that some 10 to 20 percent of their psoriasis patients are helped by antifungal medications.[8,9]

Other board certified dermatologists, including Walter B. Shelley of the Medical College of Ohio and David R. Weakley, M.D., of Dallas, also have found that psoriasis is yeast-related in many of their patients. Here are excerpts from a letter from Dr. Weakley:

> In my opinion, many cases of infectious eczematoid dermatitis are "yeast driven" and respond dramatically to antiyeast therapy. I've now seen and treated at least 30 patients who responded promptly and well to Diflucan. I've used anywhere from 200 mg. daily to 100 mg. every three days.
>
> In addition to the skin problems, a number of these patients also were troubled by migraine headaches, GI symptomatology, chronic arthritis and chronic fatigue.

My Comments

Again, the yeast, *Candida albicans,* is not **the** cause of psoriasis. Like many others, this chronic health disorder develops because of many different causes; some of which are known, and some of which aren't. Yet, if you're troubled by psoriasis, and especially if your history suggests other yeast-related problems, a several-month course of antifungal medication and a sugar-free special diet could help significantly.

Multiple Sclerosis

In 1982, Bobby Carter, a 42-year-old Jackson businessman developed numbness, tingling, muscle weakness and other symptoms. After examinations by several neurologists, he was diagnosed with multiple sclerosis. His symptoms worsened to the point that he needed a cane for support.

I saw Bobby in consultation, and because his history suggested a strong probability of yeast overgrowth, I recommended an anticandida program. After taking nystatin and nutritional supplements and following a yeast-free, sugar-free diet, Bobby improved promptly.

Through my research and the experiences of patients like Bobby, I have seen a connection between multiple sclerosis and other autoimmune disorders and yeast overgrowth.

As I said in both *The Yeast Connection* and *The Yeast Connection and the Woman:*

> Candida isn't *the* cause of MS . . . but there's growing evidence . . . based on exciting clinical experiences of many physicians that there is a yeast connection.[10]

In his observations on candida-related health problems, C. Orian Truss, M.D., described the response of a number of his patients with severe autoimmune diseases to nystatin and a low carbohydrate diet. Included were brief descriptions of several of his patients with multiple sclerosis.[11]

Like Bobby, they got better with antiyeast treatment. Bobby, for example, has continued to lead an active and productive life. In November 1994, he was elected to the Tennessee State Senate. In a recent conversation, Bobby said:

> I'm continuing my work in the Senate and I was doing great, staying on my diet, taking daily nystatin and occasional Diflucan until I was injured in a car wreck in the summer of 1998. This injury led to a definite setback with a flare-up of some of my MS symptoms . . . Skeptics might say my MS would have gotten better without the diet, nystatin and Diflucan, but I know what I'm talking about.

Support from a Cincinnati Neurologist

In January 1993, I received a letter from neurologist R. Scott Heath, M.D., who expressed an interest in the relationship of *Candida albicans* to MS. In a January 1994 conversation he said,

> *I've been impressed over the years that people with symptoms of MS who do not have spots on their brain, generally do well on antifungal treatment. Also, when people* **do** *have spots and*

we make a diagnosis of MS, they do better too. (emphasis added)

Several months after I received his letter I asked him if he would be interested in carrying out a scientific study on the yeast connection to MS. When he agreed I wrote and called the Pfizer Pharmaceutical Company. After many letters and several visits with company officials, Pfizer made a grant of $60,000 to the International Health Foundation to support a study by Dr. Heath and his colleague, Dr. Kottil Ramahan. In summarizing the findings of the study Heath said,

> In 1996, Dr. Ramahan and I completed our study of ten patients with MS treated with Diflucan. The review of our data would indicate that Diflucan does appear to show a trend in reducing the frequency of new lesions appearing in MS patients. Several of the patients in this study improved and had no new lesions on MRI.
>
> These were the patients who were the most compliant on their diet. No patients experienced clinical exacerbations of their MS while on diet and Diflucan. Our results are sufficient to study a larger group of patients over a more protracted period of time and possibly expand the study to a multi-center trial.

Here's further information about Dr. Heath's experiences, which I learned from him in December 1999:

> During the last several years I've been using the Candida Immune Complex test from AAL Reference Laboratory in Santa Ana, California, to monitor the response of my MS patients to treatment. Elevated immune complex levels in my patients seem to correlate with increased immune activity in MS and I usually treat them with Diflucan or other antifungals.
>
> If the elevated immune complex levels do not come down, I may switch the patient to Lamisil or Sporanox. Interestingly, in early MS, there is an antibody to myelin oligoendrocyte glycoprotein (MOG). Given the cross-reaction between the yeast cell wall and human proteins, these antibodies may be "looking" at the same thing. When patients have had MS for 10 or 15 years, these antibodies do not seem to be as important.
>
> If you look at the history of treatment for MS we've gone from steroids to the new beta interferon drugs. Although these

drugs may help some patients, they're not inexpensive and not without side effects. *In my practice, the anti-yeast treatment is my first step in modulating the immune system. So I almost always prescribe antifungals and dietary changes before giving Avonex, Betaseron or Copaxone.* (emphasis added)

Other Autoimmune Diseases

In his book, *The Missing Diagnosis,*[11] Dr. C. Orian Truss commented on the relationship of *Candida albicans* to multiple sclerosis, systemic lupus, erythematosus, Crohn's disease, myasthenia gravis and other autoimmune diseases.

He found that many of his patients responded favorably—and sometimes dramatically—to dietary changes and nystatin.

To gain more information about the relationship of yeast to autoimmune diseases, I interviewed a number of candida clinicians in the late 1990s. I asked them especially to comment on their experiences in treating patients who had symptoms suggestive of autoimmune diseases, yet in whom a specific diagnosis had not been made.

Here's a consensus of their comments and observations:

- Autoimmune conditions are part of a general problem of illness related to a state of immune activation. Various triggers are capable of involving the immune system in an inappropriate way. Triggers include physical and emotional injury, nutritional stress and infection. The trigger is not "the cause" of the autoimmune condition. It simply is a spark that lights the fire.
- Antiyeast medications, including nystatin, Diflucan and Sporanox, and changes in the diet can help many people with autoimmune diseases.
- Such treatment should not be regarded as a "cure," yet it helps relieve these symptoms and take the stress off the immune system.
- Some patients may show a dramatic response to antifungal therapy.

The Almost Unbelievable Story of Joyce Frederick

In August 1992, 35-year-old Joyce Frederick, who had experienced severe myasthenia gravis for 20 years, called me on the International Health Foundation Hotline.

She also told me she had taken many, many antibiotics, and that she had been troubled by recurrent vaginal yeast infections in childhood and early adolescence.

Her MG symptoms were severe and included difficulty in swallowing, loss of control of the muscles of her face, double vision, migraine headaches, and other symptoms. Although she received some help from various therapies, including Mestinon and Mytelase, she was troubled by recurring yeast infections, constipation, depression and constant eyelid weakness. Because of these symptoms, she was given additional therapies, including prednisone and gamma globulin by intravenous drip.

In spite of these therapies, Joyce continued to be troubled by PMS, headache, irritability, panic attacks, poor memory and "indescribable" body weakness and fatigue.

Because of Joyce's history of recurrent yeast infections, her improvement following dietary changes, and Dr. Truss' success in treating patients with MG using nystatin and diet, I wrote to her and said, "Call or write Dr. Truss. I feel he may be able to help you."

Early in 1993, Joyce went to Birmingham to see Dr. Truss. Following a careful history and examination, he prescribed oral and vaginal nystatin. He also stressed the importance of carefully following a restricted diet.

Within a short time, Joyce reported great improvement, including 100 percent more energy and no more 'weak' days. In March 1993, *she went on a ski vacation and skied from 7:30 in the morning to 4:30 in the afternoon!*

I've followed Joyce's progress by phone or by mail during the past seven years. Although she has had some ups and downs, she's doing well. In a December 1999 phone report to me, she said she was continuing to take nystatin and was doing "great." She can do 21 push-ups each day and recently she and her husband took a

14-day bicycle ride through the Everglades with another couple and she said, "I beat them all."

References

1. Itkin, I. H. and Dennis, M., "Bronchial hypersensitivity to extract of *Candida albicans*," *Journal of Allergy,* 1966; 37:187–195.

2. Liebeskind, A., "*Candida albicans* as an allergenic factor," *Annals of Allergy,* 394–396.

3. Gumowski, P., "Chronic asthma and rhinitis due to *Candida albicans,* epidermophyton, and tricophyton," *Annals of Allergy,* 1987; 59:48–51.

4. Ward, G. W., Hayden, M. L., Rose, G., Call, R. C., Platts-Mills, T., "Trichophyton Asthma: Reduction of Specific Bronchial Hyperreactivity Following Long Term Antifungal Therapy," *J. Allergy Clin. Immunol.,* January 1994 (abstract).

5. van der Brempt, X., Mairesse, M., and Ledent, C., "Ketoconazole (K) in Asthma: A pilot study," *J. Allergy Clin. Immunol.,* January 1994 (abstract).

6. Baker, S. M., "Nystatin for Treatment of Psoriasis." Presented at the *Candida albicans* Conference, Dallas, 1982.

7. Rosenberg, E. W., Belew, P. W., Skinner, R. B. and Crutcher, N., "Crohn's Disease and Psoriasis," (Letter/Editor) *N. Engl. J. Med.,* 1983; 308:101.

8. Crutcher, N., Rosenberg, E. W., Belew, P. W., et al, "Oral Nystatin in the Treatment of Psoriasis," *Arch. of Dermatol.,* 1984; 120:433.

9. Skinner, R. B., Rosenberg, E. W. and Noah, P. W., "Psoriasis of the Palms and Soles Is Frequently Associated with Oropharyngeal *Candida albicans,*" *ACTA Derm. Venereol* (Stockh.), 1994; Suppl. 186:149–150.

10. Crook, W. G., *The Yeast Connection and the Woman,* Professional Books, Jackson, TN, 1995; p. 222.

11. Truss, C. O., *The Missing Diagnosis,* P.O. Box 26508, Birmingham, AL 35206, 1983 and 1986; pp. 137–138.

CHAPTER · SIX

Health Problems That Affect Women More Than Men

■ ■ ■

Cystitis and Interstitial Cystitis

Cystitis

Women often enter a vicious cycle when they are treated for recurring urinary tract infections. Antibiotics, the usual method of treatment for cystitis, kill the good bacteria along with the bad. Candida yeasts aren't affected by antibiotics so they multiply to raise large families. These yeasts put out toxins that weaken the immune system, leading to further infections. Each infection is treated with antibiotics and, thus, a vicious cycle develops.

The Observations of Larrian Gillespie, M.D.

About eight years ago, I picked up a copy of Dr. Larrian Gillespie's book, *You Don't Have to Live with Cystitis.* In her comprehensive book, she emphasized the importance of urine cultures before the use of antibiotics.

> You don't want to take antibiotics if you do not have a bacterial infection . . . I steadfastly refuse to prescribe antibiotics for long periods of time except in rare exceptions . . . Why expose your entire body, as well as sensitive bladder tissue to 10 days of antibiotic therapy? . . . To me that didn't make sense.[1]

She also sharply criticized the practice of prescribing antibiotics over the phone because such practices may ultimately predispose a woman to develop interstitial cystitis.

Interstitial Cystitis (IC)

Interstitial cystitis is an inflammation beneath the membranes of the bladder. Most urologists feel that the cause is unknown and that the present standard treatments are largely ineffective.

Yet, as pointed out by Dr. Gillespie and others, when a woman has burning and frequency, she and perhaps her doctor assume she has a bacterial infection of the bladder. She may then be given an antibacterial drug.

Some patients have found, however, that their IC is helped with a special diet and antiyeast medication.

Terry Oldham, an Indiana woman with severe IC, was one of those patients. Besides having to urinate frequently and being in nearly constant severe pain, Terry also experienced bladder spasms, fatigue and mild to moderate digestive problems. She often was treated with antibiotics and, at one time, was on antibiotics for four straight months. She had to give up her job as a nurse. She said:

> My pain and fatigue were so severe I could barely stand up for more than an hour at a time. I had to urinate every 20 minutes on a bad day, and at least once an hour on a so-called "good" day . . .

Through networking Terry learned about Diflucan and Dr. Philip Mosbaugh, an Indiana urologist prescribed this antiyeast medication for her. She also avoided sugar and made other changes in her diet beginning in the summer of 1991.

Terry gradually improved over the years and in March 1998, she said,

> My bladder pain and frequency are 90% improved most of the time and I feel I have a quality of life that would not have been possible unless I'd made the changes that you and my family physician, Dr. Rick Halstead suggested.

Kay Longworth, another Indianapolis woman with interstitial cystitis, responded favorably to Diflucan and dietary changes. In February 1998, she said,

> *There's none of the burning and urgency I used to have. I go to bed about 10:00 and get up about 6:00 and I rarely have to go during the night.* I do a lot of things including diet, exercise, stress management and nutritional supplements including a teaspoon of probiotics each day. I'm experiencing a wonderful level of wellness and I want to learn more.*

In discussing IC, Dr. Gillespie said that as recently as 1979, a major urology textbook stated that it may be caused by an "emotional disorder." In responding, she said,

> Don't let them tell you it's all in your head . . . If you recognize this as the story of your life, then I would like to assure you, you don't have a psychiatric disorder. You're not crazy. Your childhood mishaps have nothing to do with the constant pain you're coping with every day of your life.

Possible Yeast Influences in Interstitial Cystitis: A Study by Philip G. Mosbaugh, M.D.

During the pást decade, this Indiana urologist has treated over 500 patients with interstitial cystitis and has served as medical adviser of the Indiana chapter of the Interstitial Cystitis Association. He also saw and followed Terry Oldham and Kay Longworth along with their family physicians. Here are excerpts from the introduction to his IC study which began in May 1997 and was completed in June 1998.**

> The goal of this study has been to evaluate whether maximum antifungal therapy (through diet, Probiotic Complex, nutritional support and oral prescription antifungal medications) has

---

*You'll find a detailed discussion of the stories of Terry Oldham and Kay Longworth in *The Yeast Connection and the Woman.*

**Dr. Mosbaugh's study was supported by the International Health Foundation with funds obtained from a grant by Pfizer Pharmaceutical Company. Physicians and other professionals who would like a copy of the 33-page protocol of this study can obtain it from the International Health Foundation, Box 3494, Jackson, TN 38303. A donation of $20 is requested.

any beneficial clinical impact on the symptom scores of selected IC patients.

Each of the 15 IC patients selected for the study gave a history of prolonged antibiotic use, elevated score on the yeast questionnaire (greater than 250 total points), failed conventional IC therapy and prior diagnosis based on standard criteria.

Upon initiation of the study, the patient followed a rigid antifungal diet for six weeks. They also took Omega 3 fatty acids (flaxseed oil), one tablespoon daily as well as vitamin/mineral complex supplements, daily during the duration of the study. After four weeks on the rigid antifungal diet, the antifungal medication, Diflucan, was administered in an initial does of 400 mg. followed by 200 mg. daily for four months.

In a May 1998 report, Dr. Mosbaugh said that six patients in the study had clearly improved, several were better in some ways and unimproved or worse in other ways; four patients dropped out of the study for various reasons. In his continuing comments, he said,

> You have to realize these people had a lot of things on their plate—bowel problems, joint aches and pains, fatigue, headache and other symptoms. It was a tough group of patients. If we do the study again, maybe we'll look at some of the patients who weren't the worst ones . . .
>
> *People with these symptoms have multiple causes and multiple therapies are necessary. Also, everybody's different. . . A lot of research is going on today and a number of medical researchers and clinicians are beginning to look at IC as a total body problem and not just one affecting the bladder.*

Sexual Dysfunction

For more than a decade, I've been hearing from tens of thousands of people seeking information and help. Some 85 to 90 percent of them are women—and most are women between 25 and 45.

One recurring complaint among the host of problems these women have suffered has been pain or discomfort during sexual intercourse (the medical term is *dyspareunia*). And many women also reported other types of sexual problems, including loss of interest and drive and inability to have an orgasm.

A Typical Letter

In 1991, a public health nurse wrote to obtain information about yeast-related problems. She expressed concerns about some of her patients, and told me her own story. Her severe PMS was causing depression and a loss of sexual desire. She wrote:

> *My decreased libido was beginning to affect my marriage. I was just not interested in sex. I had no desire for it, which really caused problems for my husband. He felt I didn't find him pleasing to me, which caused him to be depressed. The old saying goes, "When Mama's not happy, ain't nobody happy." This was so true at our house* [emphasis added].
>
> Fortunately for me, I was introduced to *The Yeast Connection.** I've been following the treatment program for almost a year now. With the diet modification, addition of vitamin supplements and nystatin, almost all of my symptoms cleared, except the vaginal candidiasis. Then I was able to obtain Diflucan, which completed the cure. I took a 200 milligram tablet for two days, and have not had a recurrence. Finally, the cycle was broken.
>
> Now at my house, everybody's happy, because Mama's happy. Our lives have most definitely been changed for the better.

A Report on One of My Patients

In March 1993, I saw 33-year-old Marjorie in consultation. Her story was typical of many women with a yeast-connected problem. All of her life she had been troubled by nose and throat infections and had taken many antibiotic drugs.

After her son was born, her health problems worsened. They included a lack of energy, PMS, yeast infections, vaginal dryness, cramps, depression, sleep problems, and poor memory. She also had a *marked loss of sex drive.* She told me:

> I saw a number of doctors who gave me different medications, including antidepressants. One doctor said, "I can't find anything

*Because this book is outdated, I do not recommend it.

physically wrong" and my tests were all negative. I know that I look terrible, feel terrible and act terrible and have absolutely no sex drive. If I didn't have such a loving husband I don't know what I would do. Please, I need help for me and my son.

Because of the severity of Marjorie's symptoms, I recommended Diflucan, a sugar-free diet and nutritional supplements. During the succeeding months, Marjorie improved. Five months later, she wrote:

> I'm feeling much better now. I'm just a new person altogether! My patience, memory, energy level, depression, sex drive and sleeping have improved 100 percent since the first time I saw you . . . But I really have to stay on my diet. I also have to exercise, which I do three or four times a week. It really helps. I can actually do a day's work without feeling "give out." My PMS is much better, but it's still there . . .

Comments by Professionals

In discussing "the clinical picture" in women with candida-related health problems in his book, *The Missing Diagnosis,* Dr. C. Orian Truss said:

> In addition to . . . symptoms that result from yeast growth in the vagina and gastrointestinal tract, numerous manifestations result from the effect on other organs, as yeast products enter the bloodstream. . . . *Diminution or total loss of sexual feeling and responsiveness is frequent.*[2]

In an article in *The Female Patient,*[3] gynecologist Jay S. Schinfeld described a study of 340 women who enrolled in the PMS unit at the University of Tennessee College of Medicine in Memphis. All were long-term PMS sufferers, he said. Many also had *significant decreases in libido.*

A number of other physicians also have told me of the frequent occurrence of sexual dysfunction in their patients with yeast-related health problems.

My Comments

Based on the observations of both professionals and nonprofessionals, sexual dysfunction (including decreased libido and orgasmic dysfunction), occurs frequently in women with yeast-related problems. And in many women, a sugar-free special diet and prescription and/or nonprescription antifungal medications may help. Yet, as I have expressed elsewhere in this book, such treatment will not be a quick fix. It should be combined with other measures, which may include nutritional supplements, control of environmental pollutants, thyroid and other endocrine therapies.

Infertility

One American couple out of every six is troubled by infertility. The causes are multiple and complex. Some are known, such as problems with sperm or Fallopian tubes blocked by endometriosis. Yet, others are unknown and a woman may fail to conceive even when sperm are active and normal, and a comprehensive infertility investigation comes up empty-handed.

As with other health problems, infertility may sometimes have a yeast connection.

More than 10 years ago, James Brodsky, M.D., a Chevy Chase, Md., internist, told me that two of his patients with infertility problems were each able to conceive within 60 days of taking nystatin and trying a special diet.

Evelyn's Story

Then I learned of a patient named Evelyn, who had married young and had a son when she was 19. She experienced no real health problems during her childhood or 20s. When she remarried in her 30s and tried to conceive again, no pregnancy resulted.

Thus began a regimen of tests and treatments for both her and her husband. About the same time, her doctor prescribed an antibiotic for a kidney infection. She began experiencing other health problems, such as increasingly severe dysmenorrhea, lower back pain and general fatigue. She was diagnosed with mild to moderate endometriosis, although another test showed clear tubes.

She was given more antibiotics for other infections. At the same time, the cause of her pain couldn't be found and she still wasn't pregnant. Then she read *The Yeast Connection* and took the quiz. She said:

> My score was 181, which indicated the "candida almost certainly was playing a role in causing my health problems." Among the items in my history which added significantly to my score were repeated courses of antibiotics, birth control pills, worsened symptoms on damp days and sugar craving. Also, fatigue, the feeling of being drained, depression, abdominal pain, and PMS.

A physician consultant interested in yeast-related health problems told her that although candida was not the cause of her reproductive problems, it could be playing a significant role. He suggested a therapeutic trial of nystatin, diet and nutritional supplements.

After she had been on the program four weeks, she called herself a new woman. And she said, "my depression, fatigue, bloating, mood swings, urinary symptoms and vaginal burning had all but vanished."

Then came the best news: She was pregnant. Nine months and three weeks after starting on nystatin, she and her husband became new parents.

Other Comments

Other physicians have told me about infertility patients who got pregnant after anti-candida therapy.

Gynecologist Richard Mabray, M.D., of Victoria, Texas, for example, considers *Candida albicans* an important part of many women's problems, including infertility. It's not the cause, he said, but "a major trigger and part of the vicious cycle that appears to start with immune dysfunction." In summarizing his experiences, Dr. Mabray said:

> I cannot base my life and practice on "the yeast connection." However, this is an extremely common problem with interaction and multiple facets for a large portion of my patients. Therefore, even though it is not the total answer, it would be impossible to practice good medicine without this understanding. . . . I simply would fail to meet the needs of a large proportion of my practice.

My Comments

Yeast overgrowth is certainly not "the cause" of infertility. Yet, for a couple struggling in vain to solve the problem, anti-candida therapy is a safe option worth considering. Such therapy should be comprehensive and should feature oral nystatin and a sugar-free special diet. I would especially recommend such a treatment program if either or both partners gave the typical history.

References

1. Gillespie, L., *You Don't Have to Live with Cystitis*, Avon Books, 1986.

2. Truss, C. O., *The Missing Diagnosis*, Birmingham, AL, 1986; pp. 37–42.

3. Schinfeld, J., "PMS and Candidiasis: Study explores possible link," *The Female Patient*, July 1987; 12:66–69.

CHAPTER · SEVEN

Yeast-Related Problems That Affect Only Women

■ ■ ■

Recurrent Vaginal Yeast Infections

About four years ago, the vagina—an area of a woman's anatomy that had rarely been talked about in public—became front page news. A major factor in prompting this publicity was a decision by the government to allow women to buy antiyeast, vaginal suppositories over the counter.

Some women are rarely—if ever—troubled by a yeast infection. Other women experience these infections time after time. One reason is that the common yeast, *Candida albicans,* thrives in a dark, warm environment. And the vagina serves as an ideal place for this yeast to live and multiply—especially when other factors encourage its growth. These factors include:

- **Antibiotics.** These drugs, including amoxicillin, ampicillin, Ceclor, Keflex, Augmentin and tetracycline, have been looked upon as "wonder drugs" and are freely used in both people and animals. They do save lives by eradicating germs that cause serious infections, including pneumonia and meningitis. Antibiotics, however, may cause more problems than they solve when they are given unnecessarily and for too long. Among these problems are recurrent vaginal yeast infections.
- **A weak immune system.** A Japanese researcher in the 1970s published studies showing that Candida may act as an immunosuppressant.[1] During the 1980s, a Cornell University

researcher published an article stating that antibiotic related alterations in the flora of the digestive tract may adversely affect the immune system and lead to repeated vaginal yeast infections.[2]

- **Overgrowth of yeast in the digestive tract.** In the 1970s, Michigan State University researchers studied 98 young women who were troubled with recurrent vaginitis. After finding that all of these women had yeast in their stools, they concluded that a cure for vaginitis is not possible without controlling yeast in the digestive tract.[3]
- **Diet.** A number of research studies during the past 10 years showed that diets high in sugar make women much more apt to develop a vaginal yeast infection.[4,5,6]

A 1993 study at St. Jude Children's Research Hospital in Memphis documented the role sugar plays in encouraging yeast overgrowth in the intestinal tract. One group of mice was given plain water and a second group was given water containing dextrose. *The results showed that the growth of* Candida albicans *in the digestive tract was 200 times greater in the mice receiving dextrose than in those who received only water.*[7]

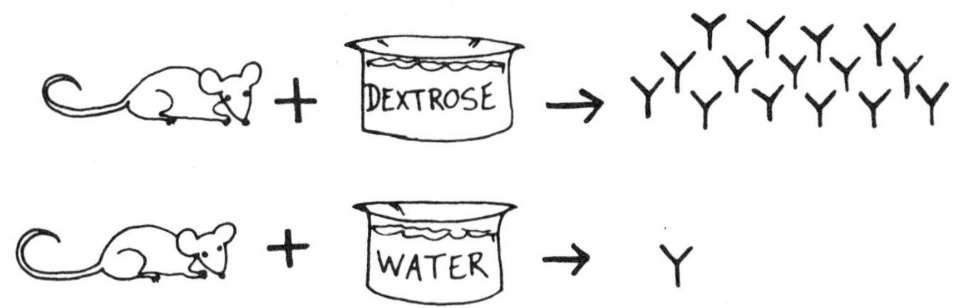

Overcoming Vaginal Yeast Infections

If you continue to be bothered by vaginitis, here are my recommendations:

- First go to your gynecologist or family physician to make sure that you do not have an infection caused by a germ other than *Candida albicans.*
- Change your diet. Eat more vegetables and other nutritious foods and cut way back on your sugar intake. Take sugar-free,

fruit-free yogurt, and/or high quality probiotics. These substances contain friendly bacteria, including *Lactobacillus acidophilus* and *Bifidobacterium bifidum*—the same ones found in yogurt. You can obtain them in capsules or powders from a health food store or pharmacy.

- Take oral antiyeast medications. These include caprylic acid products, citrus seed extracts and other non-prescription products that you can obtain at a health food store. Or, you can ask your physician or nurse practitioner to prescribe oral antifungal medications, including nystatin or Diflucan. Most medications may need to be taken for many weeks.
- Candida immunotherapy may help.[8]

(You'll find more information in chapters 9 and 10.)

Vulvodynia (Vulvar Burning)

First coined in 1978, vulvodynia is defined as chronic vulvar discomfort, characterized by burning, stinging, irritation or rawness.

During the past decade, a number of professionals, including Atlanta gynecologist/dermatologist, Marilynne McKay, M.D., have studied this often devastating and frustrating disorder. Again, an excessive amount of yeast may be part of the problem.

In a comprehensive article published in 1992, Dr. McKay said,

> Bacterial, fungal, and viral infections should all be considered in this area, occurring as either primary or secondary problems. . . . Candida is by far the most important infectious agent to consider in evaluation of patients with vulvodynia.[9]

In a phone conversation with me she told me she treated all of her vulvodynia patients with low dose antifungal therapy for four to six months and that she found Diflucan 100 milligrams, once or twice a week quite effective.

Vulvodynia, like many other chronic health disorders, including PMS, endometriosis, chronic fatigue syndrome and interstitial cystitis, is rarely, if ever, due to a single cause. Vulvodynia, for example, also can be caused by excessive oxalate in the urine. This irritating material, which has long been known to cause pain, is produced in the body during normal metabolism. It causes pain when it comes in contact with nerve endings. In some women, however, the cause of vulvodynia is unknown.

Suggestions for Treating Vulvodynia

- First, try a sugar-free diet and ask your physician to prescribe Diflucan for a month. This treatment plan is especially indicated in women who have a history of taking antibiotics.
- If the pain is not relieved with antiyeast treatment, it may respond to a treatment program that lessens the excretion of oxalate in the urine.[10]
- Dr. McKay also has found that a low dose of amitriptyline may help relieve your burning pain and related symptoms.[11]
- For more information contact the National Vulvodynia Association, P.O. Box 4491, Silver Spring, MD 29014-4491; Phone (301) 299-0775; FAX (301) 299-3999.

Premenstrual Syndrome (PMS)

Several times each week, the average American, both male and female, hears or reads about PMS. Frequent newspaper and magazine articles and programs and advertisements on TV have now made PMS a household word.

For many women, PMS is real. They're not imagining the mood swings, the headaches, fatigue, breast tenderness and other symptoms. As with most problems, PMS is caused by multiple factors and has multiple treatments. However, many professionals are seeing a link between yeast and PMS. And many patients have seen their lives changed when they try a sugar-free special diet, oral antiyeast medications and nutritional supplements, especially Vitamin B_6.

More than 15 years ago, a California gynecologist, Dr. Guy E. Abraham,* started publishing his observations that many women

*Dr. Abraham has written a number of other articles and several booklets, which document the role nutrition plays in women's health problems, including PMS and post menopausal osteoporosis. Information can be obtained by writing Guy E. Abraham, M.D., F.A.C.N., 2720 Monterey Street, Suite 406, Torrance, CA 90503.

with PMS showed significant nutritional deficiencies.[12] And he found that by supplementing his patients with magnesium, vitamin B_6 and other nutrients, they improved significantly. Abraham outlined four types of PMS:

- The most common type of PMS caused premenstrual anxiety, irritability and nervous tension. Women in this subgroup consume an excessive amount of dairy products and refined sugar.
- The second most common type was associated with symptoms of water and salt retention, abdominal bloating, breast pain and weight gain.
- A third group with PMS had premenstrual craving for sweets, increased appetite and indulgence in eating refined sugar. This is followed by palpitations, fatigue, fainting spells, headaches and sometimes the shakes.
- A fourth group of PMS patients, the least common type, had the dangerous symptoms of depression, withdrawal, insomnia, forgetfulness and confusion. Suicide was most frequent with this group.

The Yeast Connection to PMS

In his superb book, *The Missing Diagnosis,* Dr. C. Orian Truss links PMS symptoms to yeast problems. He observed the favorable response of women with PMS to antifungal medication and diet.[13] In my own practice, I confirmed Dr. Truss' observations with many of my patients. I saw dozens of women who complained of irritability, fatigue, depression and other symptoms that became worse the week before menstruation.

One of my patients was a mother and wife named Arlene. Over several years her premenstrual problems had worsened to the point that she had severe mood swings during the five or six days before her period. She told me:

> Increasingly, I began to have "attacks" where I would get so upset I would clinch my fists, grit my teeth, wring my hands and tense every muscle in my body. I'd feel like screaming. I wouldn't be able to sit down or lie down. I would usually end up

> getting in the car and driving until I calmed down . . . Even when I wasn't having attacks, I found I could not cope without getting upset with any situation. Such as the washer or dryer going out, or my husband forgetting to call and tell me he'd be late coming home after work.

Arlene told me that before the development of her severe PMS she had experienced many urinary tract infections and had to take a lot of antibiotics. I first sent Arlene to a gynecologist for a careful checkup. Then I put her on a sugar-free special diet and nystatin. Within three weeks she was much better. Seven weeks after our first visit she told me:

> For the first time in three years my premenstrual period wasn't hell.

In a 1985–86 study at the University of Tennessee, Jay S. Schinfeld, M.D.,* studied the effect of using oral anticandida agents and yeast elimination diets to treat women with severe PMS and a history of vaginal candidiasis for whom standardized therapy had failed. In an article published in *The Female Patient,* Schinfeld reported that treated patients showed significant physical and psychological improvement over untreated controls.[14]

A number of professionals, including John Curlin, M.D., Carol Jessop, M.D., George Miller, M.D., Richard Mabray, M.D., Jean Rowe, R.N., Susanna Choi, M.D. and Kathy Gibbons, Ph.D. have noted that most of their PMS patients improved significantly—and sometimes dramatically on a treatment program which features antifungal therapy and dietary changes.

More About Nutritional Factors and PMS

In his excellent book, *Healing Through Nutrition,*[15] Dr. Melvyn Werbach provides readers with a comprehensive discussion of the many dietary factors that play a role in causing anxiety, irritability, sugar craving, headache and other symptoms experienced by women with PMS.

---

*Dr. Schinfeld is now an associate professor, Department of Obstetrics and Gynecology, Temple University School of Medicine.

In another book, *PMS—Solving the Puzzle*,[16] Linaya Hahn discusses the importance of dietary changes. She targets caffeine as a major troublemaker.

In her discussion, Hahn points out that beverages with caffeine, when combined with sugar, add to the problem. She suggests that PMS sufferers go off caffeine gradually by cutting consumption by one-third for a few days and then another one-third for a few days and then eliminate the final one-third. She also gives readers some easy-to-follow practical suggestions, along with a list of foods to eat and foods to avoid.

Endometriosis

Endometriosis is a misunderstood disease that affects an estimated five million women in this country alone. Much of my understanding about this disease has come from Mary Lou Ballweg, founder and head of the Endometriosis Association, a self-help organization.

More About Mary Lou Ballweg and the Endometriosis Association (EA)

Dedicated, caring, persistent people often can advance the frontiers of medicine more than physicians and other "experts." Ballweg is one of those people.

In her ten-page introduction to *Overcoming Endometriosis*, Ballweg explained that in 1978, she was a healthy, productive businesswoman. Then, after a series of health problems, she found out she had endometriosis. None of a series of different treatments were really satisfactory. Ballweg said:

> I decided only a group of other women who had "been there" were likely to understand and provide the emotional support I needed to come to terms with this baffling experience of endometriosis.

So, in the fall of 1979, with the help and encouragement of a number of other people, including Carolyn Keith, Fran Kaplan and Dr. Karen Lamb, the Endometriosis Association was born. The association worked hard at dispelling the "nightmare of misinformation" surrounding the painful, chronic and stubborn disease. One such myth is that endometriosis is a disease of white, stressed out, upper socioeconomic women who bring it on themselves by postponing childbearing. Said Ballweg:

> Only when the Endometriosis Association began in 1980 and systematically gathered data were we able to disprove all of these myths . . . Endometriosis is, in fact, an equal opportunity disease affecting all races, personalities, socioeconomic groups, as well as all ages of females from as young as 10 or 11 to as old as women in their 60s and 70s.

Ballweg called endometriosis just the tip of the iceberg for a whole range of health problems that are related to hormonal/immune dysregulation. Besides the traditional symptoms of chronic pelvic pain, infertility, pain with sex and gastrointestinal problems, Ballweg said, women with endometriosis also are more apt to be troubled by a host of other problems. These include:

- asthma and eczema
- food intolerances
- chemical sensitivities
- mitral valve prolapse
- a tendency to infections
- mononucleosis
- chronic fatigue syndrome (CFS)
- fibromyalgia
- autoimmune disorders, including lupus and Hashimoto's thyroiditis

I also learned from Ballweg that toxic chemicals play a major role in causing endometriosis. In a study by the Endometriosis Association, 79 percent of a group of animals exposed to dioxin developed endometriosis.

The Yeast Connection to Endometriosis

Ballweg also told me that many members of the Endometriosis Association found that nystatin, Nizoral or Diflucan and a sugar-free special diet helped them overcome many of their symptoms. And in the *Endometriosis Association* newsletter, I read the stories of several women with endometriosis whose symptoms were improved following dietary changes and antifungal medication.

Professionals who have found the connection between endometriosis and yeast include Wayne H. Konetzki, a Wakashau, Wisconsin, allergist and Arnold Kresch, M.D., a Palo Alto, California, gynecologist. Each of these professionals have found that they could help their patients with oral antifungal medication, a special diet and nutritional supplements. In a 1995 meeting of the Endometriosis Association, they reported the favorable response of many of their endometriosis patients to immune therapy using tiny sublingual doses of *Candida albicans* extract.

Further evidence of the relationship of *Candida albicans* to endometriosis was published in a January 1997 article in Patient Care entitled, "Current Approaches to Endometriosis." Under the heading, "A Budding Treatment for Endometriosis," appeared these comments.

> Immune therapy is a multi-faceted approach to endometriosis treatment and relies on a technique developed more than 4,000 years ago in China: Oral Tolerization. To identify hormones a patient is allergic to and help determine the doses . . . clinicians experienced in immune therapy administer an intradermal skin test.
>
> Patients also have a sublingual test for an allergic reaction to *Candida albicans* . . . If the endometriosis symptoms worsen (after receiving a small drop of *Candida albicans* under their tongue), the patient is considered allergic to the organism and in need of treatment that includes Oral Tolerization, antifungal drugs and a proper diet.

My Comments

Obviously candida isn't the cause of endometriosis,* PMS and the many other health problems that affect women. Yet, based on reports I've received from both professionals and nonprofessionals,

---

*For more information about endometriosis, including education, support and research, write to the Endometriosis Association, International Headquarters, 8585 N. 76th Place, Milwaukee, WI 53223. 1-800-992-3636 or 414-355-2200. Fax 414-355-6065. This organization, which has chapters throughout the world, carries out medical research; publishes fact sheets, brochures and a newsletter; serves as an information clearinghouse; and provides technical assistance. See also pages 258–259.

a sugar-free special diet and antifungal medication often help women with these distressing disorders.

References

1. Iwata, K. and Uchida, K., "Cellular Immunity in Experimental Fungal Infections in Mice," *1978 Mykosen, Suppl.* 1, 72–81.

2. Witkin, S. S., "Defective Immune Responses in Patients with Recurrent Candidiasis," *Infections in Medicine,* May/June, 1985; pp. 129–132.

3. Miles, M. R., Olsen, L. and Rogers, A., "Recurrent Vaginal Candidiasis: Importance of an Intestinal Reservoir," *JAMA,* 238:1836–1837, October 28, 1977.

4. Reed, B. D., and associates, "The Association Between Dietary Intake and History of Candida Vulvogatinitis," *J. of Family Practice,* 1989; 29:509–515.

5. Horowitz, B. J., Edelstein, S., and Lippman, L., "Sugar Chromatography Studies in Recurrent Candida Vulvovaginitis," *J. Reproduct. Med.,* 1984; 29:441–443.

6. Gilmore, B. J., et al, "An ic3b Receptor on *Candida albicans:* Structure, function and correlates for pathogenicity," *Journal of Infectious Diseases,* 1988; 157:38–46.

7. Vargas, S. L., Patrick, C. C., Ayers, G. D. and Hughes, W. T., "Modulating Effect of Dietary Carbohydrate Supplementation on *Candida albicans,* Colonization and Invasion in a Neutropenic Mouse Model," *Infection and Immunity,* February 1993; 61:619–626.

8. Crandall, M.: *J. Advancement in Med.,* 1991; 4:21–38.

9. McKay, M., "VULVODYNIA Diagnostic Patterns," *Dermatol. Clin.,* 1992; 10:423–433.

10. For more information about this treatment program and *The Low Oxalate Cookbook,* write to the Vulvar Pain Foundation, P.O. Drawer 177, Graham, NC 27253.

11. McKay, M., "Dysesthetic [Essential] Vulvodynia Treatment with Amitriptyline," *J. Reprod. Med.,* 1993; 38:9–13.

12. Abraham, G. E., "Nutritional Factors in the Etiology of Premenstrual Tension Syndrome," *J. Reprod. Med.,* 1983; 28:446.

13. Truss, C. O., *The Missing Diagnosis,* P.O. Box 26508, Birmingham, AL 35226, 1986; pp. 19–31.

14. Schinfeld, J. S., "PMS and Candidiasis: Study Explores Possible Link," *The Female Patient,* 1987; 12:66–69.

15. Werbach, M., *Healing Through Nutrition,* HarperCollins, New York, 1993; pp. 325–331.

16. Hahn, L., *PMS: Solving the Puzzle,* Chicago Spectrum Press, Evanston, IL, 1995; pp. 43–44.

CHAPTER · EIGHT

Children's Health Problems

▪ ▪ ▪

As a pediatrician I'm interested in problems that affect millions of infants, young children and teenagers. I'm especially concerned about recurrent ear disorders, behavior and learning problems and juvenile delinquency. Moreover, reports in the medical and lay literature show . . .

- These problems are increasing in frequency and are related.
- Present methods of management are ineffective.
- These problems are often yeast-related.

Yeast Questionnaire for Children

Filling out and scoring this questionnaire should help you and your physician evaluate the role Candida albicans *contributes to your child's health problems.* Circle the appropriate point score for questions you answer "yes." Total your score and record it at the end of the questionnaire.

	Point Score
1. During the two years before your child was born, were you bothered by recurrent vaginitis, menstrual irregularities, premenstrual tension, fatigue, headache, depression, digestive disorders or "feeling bad all over?"	30

Yeast Questionnaire for Children—*Cont'd.*

	Point Score		
2. Was your child bothered by thrush? (Score 10 if mild, score 20 if severe.)	10	20	
3. Was your child bothered by frequent diaper rashes in infancy? (Score 10 if mild, 20 if severe or persistent)	10	20	
4. During infancy, was your child bothered by colic and irritability lasting over 3 months? (Score 10 if mild, 20 if moderate to severe.)	10	20	
5. Are his symptoms worse on damp days or in damp or moldy places?		20	
6. Has your child been bothered by recurrent or persistent "athlete's foot" or chronic fungous infections of his skin or nails?			30
7. Has your child been bothered by recurrent hives, eczema or other skin problems?	10		
8. Has your child received: a. 4 or more courses of antibiotic drugs during the past year? Or has he received continuous "prophylactic" courses of antibiotic drugs?			60
b. 8 or more courses of "broad-spectrum" antibiotics (such as amoxicillin, Keflex, Septra, Bactrim or Ceclor) during the past three years?			40
9. Has your child experienced recurrent ear problems?		20	
10. Has your child had tubes inserted in his ears?	10		
11. Has your child been labeled "hyperactive"? (Score 10 if mild, 20 if moderate to severe.)	10	20	
12. Is your child bothered by learning problems (even though his early development history was normal)?	10		
13. Does your child have a short attention span?	10		
14. Is your child persistently irritable, unhappy and hard to please?	10		

	Point Score		
15. Has your child been bothered by persistent or recurrent digestive problems, including constipation, diarrhea, bloating or excessive gas? (Score 10 if mild; 20 if moderate; 30 if severe)	10	20	30
16. Has he been bothered by persistent nasal congestion, cough and/or wheezing?	10		
17. Is your child unusually tired or unhappy or depressed? (Score 10 if mild, 20 if severe.)	10	20	
18. Has he been bothered by recurrent headaches, abdominal pain, or muscle aches? (Score 10 if mild, 20 if severe)	10	20	
19. Does your child crave sweets?	10		
20. Does exposure to perfume, insecticides, gas or other chemicals provoke moderate to severe symptoms?			30
21. Does tobacco smoke really bother him?		20	
22. Do you feel that your child isn't well, yet diagnostic tests and studies haven't revealed the cause?	10		
TOTAL SCORE			

Yeasts possibly play a role in causing health problems in children with scores of 60 or more.

Yeasts probably play a role in causing health problems in children with scores of 100 or more.

Yeasts almost certainly play a role in causing health problems in children with scores of 140 or more.

Ear Disorders

In his book, *The Missing Diagnosis,* C. Orian Truss, M.D., told the story of a 16-month-old boy who gave a history of almost constant health problems beginning at the age of two-and-a-half months. At that time, the infant was given his first antibiotic for a cold. During ensuing months, he was given antibiotics repeatedly for ear and other infections. And at the age of 10 months, tubes were put in both ears.

At age one, because of the child's continued respiratory problems, his parents brought him to Dr. Truss, who prescribed oral nystatin and dietary changes. After four months of this treatment, the child's irritability, diarrhea and ear problems had vanished. In discussing his experiences in treating this child, Dr. Truss said,

> In my opinion, this is not an isolated problem. In fact, it is probably very common. Antibiotics save countless lives, but as with most forms of medical treatment, some individuals are left with residual problems related to their use.
>
> Perhaps a single most fascinating and potentially important aspect of this case was the abrupt cessation of the ear infections. This suggests that *Candida albicans* was actually causing this problem and makes one wonder about the possible relationship of this yeast to what seems almost a national epidemic of otitis and "tubes in the ears."

Truss also discussed other health problems in children, including learning disabilities and depression. He said:

> At any age, but particularly in young children who experience difficulty with school, this condition (candidiasis) is worth considering.[1]

In 1987, Kenneth Grundfast, M.D., chairman of the Department of Otolaryngology of Children's Hospital National Medical Center in Washington, D.C., and Cynthia J. Carney, a health writer and former contributor to the *Washington Post* health section, published a book, *Ear Infections in Your Child.*[2]

I was pleased to see that they pointed out that allergies to milk and other foods, and the frequent and/or indiscriminate use of antibiotics may contribute to recurrent problems. And I was even more pleased—even excited—to read their brief comments about fungal infections in children who had been treated with prophylactic antibiotics and to see their reference to *The Yeast Connection.*

Ear Problems and Food Allergies

Middle ear problems can occur in children who are *not* allergic. Yet, over thirty years ago, Dr. John P. McGovern and associates said that these problems occur with far greater frequency in children who suffer from an allergic swelling of the linings of the nose and throat.

This prominent Texas physician and his associates studied 512 children with allergy and ear trouble. They reported their findings in the April 10, 1967 issue of the *Journal of the American Medical Association.* They found that following careful allergy study and management, 97% of these patients had good or excellent results.

Even before Dr. McGovern's study, a number of ear, nose and throat specialists had discussed the relationship of allergies—including food allergies—to ear disorders. And at a 1982 meeting of the Academy of Environmental Medicine in New York, Dr. George E. Shambaugh, Professor Emeritus of Otolaryngology at Northwestern University, gave an address entitled, "Serous Otitis: Are Tubes the Answer?"

In this address, Dr. Shambaugh commented, "*Serous otitis*—is the largest single cause of hearing loss in children. And the operation of inserting a ventilating tube through the tympanic membrane (ear drum) to restore hearing loss has become the most frequent hospital-surgical procedure with anesthesia today."

Dr. Shambaugh asked pediatricians and other physicians to take a look at the allergic aspects of ear problems in children. And he pointed out that a program of allergic management, including attention to *hidden* or *delayed onset* food allergy, helped him manage recurrent ear problems in many children. He said that his results with allergic management were far better than those obtained by putting children on prolonged courses of antibiotics or relying on tubes to clear up the condition.

In spite of these observations by Dr. Shambaugh and many other physicians, the role of food allergy in causing ear problems received little attention through the 1980s and on into the 90s. Then, T. M. Nsouli, M.D., and colleagues at Georgetown University published an article entitled, "Role of food allergy in serous otitus media" in the September 1994 issue of the *Annals of Allergy.*

In summarizing their results, these investigators found that there was a significant statistical association—between food allergy and recurrent serous otitis in 81/104 patients (78%). The elimination diet led to a significant amelioration of serous otitis media in 70/81 patients as assessed by clinical evaluation and tympanometry. The challenge diet with the suspected offending food(s) provoked a recurrence of serous otitis media in 66/70 patients (94%).

And in their concluding statement they said, *"The possibility of food allergy should be considered in all pediatric patients with recurrent serous otitis media and a diligent search for the putative food allergen made for proper diagnostic and therapeutic intervention."* (Emphasis added.)

Ear Problems and Yeast Overgrowth

In my continuing efforts to bring the relationship of yeast overgrowth to ear infections into the medical mainstream, I wrote a letter to the editor of the *Ear, Nose and Throat Journal,* which was published in the May 1992 issue. In my letter, I suggested a simple study, which I hoped would provide information that would help interrupt the vicious cycle of ear infections in young infants. In 1996, I discussed ways of carrying out this study with several consultants. Here's what we came up with:

- Physicians who see newborns would be recruited to participate in the study. Each physician would keep records on 15 consecutive newborns who would be seen on a regular basis during the first year.
- Two physicians would participate in the initial study (I'll call them Physician A and Physician B).
- Physician A would follow the infants according to his usual practice (this would be the control group).
- Physician B would prescribe oral nystatin powder (250,000 units) three times a day for infants who were receiving broad-spectrum antibiotics. The nystatin would be continued for three weeks after the antibiotic was discontinued.

  Infants followed by Physician B would be given ¼ teaspoon of VitalDophilus powder along with each dose of nystatin powder. This probiotic helps maintain a normal balance of friendly bacteria in the digestive tract. Infants who experienced a second ear infection would be continued on the same doses of nystatin and probiotics twice daily for four months.

At this time (summer 1998), no physician has stepped forward to carry out the study on the relationship of repeated antibiotics and yeast overgrowth to ear problems. I continue to feel that such a study is relevant. Pediatricians or family physicians who are interested in carrying out a study can obtain more information by writing to the International Health Foundation, Box 3494, Jackson, TN 38303.

In February 1997, an article in the *Journal of Allergy & Immunology* provides additional, but indirect, support to the hypothesis which was first proposed by Dr. C. Orian Truss over a decade ago (see page 74). This study, carried out by researchers in Finland in a double-blind study, showed that when probiotics were used in infants with food allergies that children were improved significantly compared to controls. (See also page 11.)

Ear Problems in Infancy and Later Hyperactivity

In August 1982, Wesley, a two-and-a-half-year-old boy, was referred to me for an evaluation because of severe nervous symptoms, including temper tantrums and hyperactivity. In reviewing his history, I learned that he had suffered repeated ear problems, which were treated with antibiotics. During one two-month period, he was given antibiotics every day in an effort to "suppress" the ear infections.

At age two, because of his temper tantrums and other severe nervous system symptoms, Wesley's pediatrician referred him to a clinical psychologist who advised behavior modification. Yet, it didn't work and the behavior problems continued. So did the bouts of ear infections.

At the time of my first visit, Wesley's mother told me that sweets of any kind triggered behavioral symptoms. Corn also was a major troublemaker.

Because of these symptoms, and the history of multiple courses of antibiotics, I prescribed oral nystatin powder and a sugar-free, corn-free diet. In one month, Wesley was "like a different child"; yet, when challenged with sugar and junk food, the hyperactivity and irritability returned.

Wesley continued the diet and nystatin on a regular basis for two years. Then he was able to relax a bit on the diet. Major infractions, however, would always cause problems.

In the Fall of 1995, I called Wesley's mother to find out how he was getting along. She told me he was 16 years old, ate wholesome foods, and took vitamins, magnesium and occasional nystatin. She said, "He never had to take Ritalin or other drugs. He's doing well in self-esteem and in his school work."

More About Ear Problems in Infancy and Later Hyperactivity

In May 1987, researchers from the University of Colorado and Yeshiva University published a study showing that infants who had experienced repeated bouts of ear problems were prone to develop hyperactivity later. Here's an excerpt from the abstract of their studies:

> An association between the frequency of otitis media in early childhood, and later hyperactivity, is reported in this study. The subjects were 67 children referred to a child development clinic for evaluation of school failure. Ranging from six to thirteen years old, all the children demonstrated specific learning problems, and 27 were considered hyperactive by two or more raters.
>
> Sixteen of the hyperactive children were treated with central nervous system stimulant medication. In retrospect, *there was positive correlation between an increasing number of otitis media infections in early childhood and the presence and severity of hyperactive behavior.*

These observers found that 69 percent of children medicated for hyperactivity gave a history of more than ten ear infections. By comparison, only 20 percent of nonhyperactive school failure patients had a similar number of infections. Also, 22 out of the 28 children who gave a history of more than 10 infections experienced repeated ear problems before the age of one year.

In discussing their findings, these authors pointed out that not all hyperactive children give a history of early otitis and not all children with recurrent otitis become hyperactive. They speculated on possible reasons for their findings and said:

> Further investigation is necessary to evaluate etiologic aspects of this association.[3]

In my opinion, repeated antibiotics given for ear infections set up a vicious cycle, which includes recurrent infections and nervous symptoms of various types. Here's a chart which summarizes what I feel is happening:

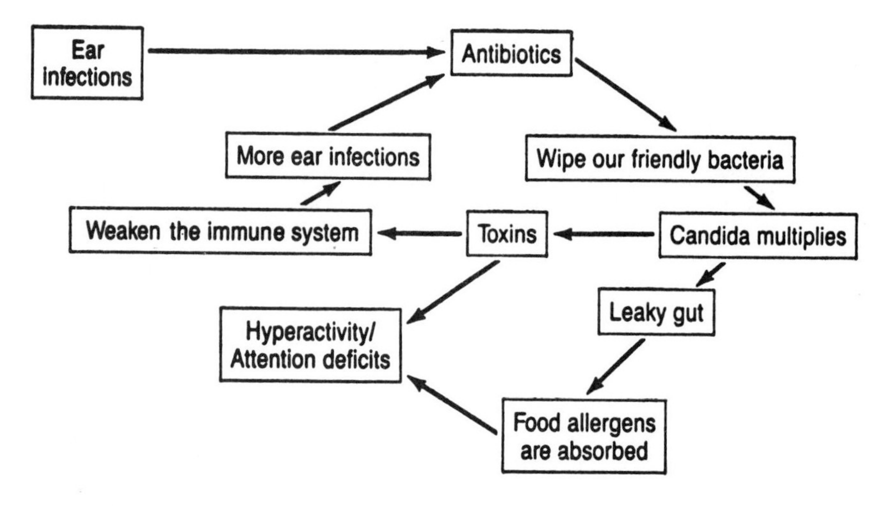

Hyperactivity and the Attention Deficit Disorder (ADHD)

Observations in My Own Practice

Although I didn't know that systemic and nervous symptoms in my pediatric patients could be yeast related, beginning in the late 1950s I learned that such symptoms were often caused by food sensitivities. Common troublemakers included especially milk, wheat, corn, chocolate and egg.

Yet, it wasn't until the early '70s that I learned that the ingestion of cane, beet or corn sugar was an important cause of hyperactivity—even though I didn't understand the mechanism.

In the 1970s, at the suggestion of the late Dr. Amos Christie (my pediatric chief at Vanderbilt), I kept a record of every new patient who came to see me because of complaints of hyperactivity, attention deficits and other behavior and learning problems.

During a five-year period (January 1, 1973—December 31, 1977), I saw 182 children with these complaints. Most were between four and eight years old.

Each child was given a comprehensive work-up which included a carefully planned and executed seven-day elimination diet. The eliminated foods included beet- and cane-sugar, milk, wheat, corn, egg, chocolate, yeast, citrus and food coloring and additives. If and when the child's symptoms improved (as they usually did), the eliminated foods were returned to the diet one food per day and reactions were noted. At the end of five years, using questionnaires and interviews, I obtained the following information:

- Seventy percent (128) of the parents were absolutely certain that their child's hyperactivity was diet related.
- The foods causing hyperactivity in 136 children included sugar, 77; colors, additives and flavors (especially red food coloring), 48; milk, 38; corn, 30; chocolate, 28; and wheat, 15.
- Many other foods were reported as causing trouble.

I published my observations in an article entitled, "Can What a Child Eats Make Him Dull, Stupid or Hyperactive?"[4]

Scientific Support for the Relationship of Diet to ADHD

Observations of Others

During the mid-1980s, three reports in major medical journals documented the importance of dietary changes in helping children with ADHD and other neurological problems.[5, 6, 7]

Using a few-foods diet in studying the children with attention deficit disorder, British researchers stated,

> The results of a crossover trial on 19 children showed a significant effect for the provoking foods to worsen ratings of behavior and to impair psychological test performance. This study shows that observations of changes in behavior associated with diet made by parents . . . can be reproduced using double-blind methodology and objective assessments.[8]

Australian researchers published observations on the relationship of diet to behavior in an American pediatric journal. Using a 21-day double-blind, placebo-controlled study, they found that behavioral changes and irritability, restlessness and sleep disturbances, were associated with the ingestion of food dyes in some children.[9]

In 1994, American researchers also documented the relationship of diet to neurological problems in children. Here's a brief excerpt from their study:

> Nineteen of 26 children with ADHD (73%) responded favorably to an elimination diet (P<.001) . . . a double-blind, placebo-controlled food challenge was completed in 16 children.[10]

Should Ritalin Be the Main Therapy for Children with ADHD

In spite of these studies and the observations of hundreds of physicians and thousands of parents, most professionals have ignored the relationship of diet to ADHD and have relied mainly on Ritalin. For example, in a cover story article, "Medical Experts Defend Against Ritalin Charges," in the May 1996 issue of *AAP News* (the official newspaper of the American Academy of Pediatrics—AAP), three "medical experts" defended the increasing use of Ritalin.

In responding to the charge by the International Narcotics Control Board (INCB) that Ritalin is overprescribed and used 10 times more in the U.S. than in the rest of the world, one of these experts said:

> The INCB should even be more worried that other countries are not prescribing more . . . that they're not taking advantage of this treatment and are writing off children who are not succeeding in life.

The *AAP News* article prompted me to respond and my comments were published in the July 1996 issue of *AAP News*. Here are excerpts,

> The incidence of ADHD has increased sharply during the past 40 years. I'm certain that this is an actual increase and not due to increased recognition. . . . Food sensitivities and other dietary factors are clearly related to ADHD.
>
> I urge the American Academy of Pediatrics to host a Workshop/Think Tank on ADHD, its causes and management. Participants should include pediatricians and other professionals and

non-professionals with different experiences and persuasions. The lives of countless children and their families are at stake.

During the next two-and-a-half years I continued my "crusade" to bring the relationship of diet, food allergies/sensitivities, environmental toxins, yeast overgrowth and other causes of ADHD into the medical and public mainstream. And in December 1998 I learned that Joseph A. Bellanti, a Georgetown University pediatric allergist and Richard E. Layton, a Towson, Maryland pediatrician, shared my interests and concerns.

The November 1999 conference, *ADHD: Causes and Possible Solutions*

In early 1999, Bellanti, Layton and I began planning a conference on children with ADHD and related disorders. Our goal was to include participants who would present a balanced view of conventional and nonconventional information. The conference, *ADHD: Causes and Possible Solutions,* was held in Arlington, Virginia, November 4–7, 1999. It was sponsored by Georgetown University Medical Center and co-sponsored by the International Health Foundation (IHF) and the International Center for Interdisciplinary Studies of Immunology (ICISI) at Georgetown University Medical Center.

Conference participants included Dr. Marianne Glanzman (University of Pennsylvania) who asked, 'What is ADHD?" and discussed various causes; Dr. Donald R. Davis (University of Texas) and Dr. Leo Galland (New York) who presented data which showed that *most American children are consuming inadequate amounts of essential nutrients;* Dr. Stephen J. Schoenthaler (University of California) who described the impact of poor nutrition on violence, delinquency, intelligence, and academic performance.

Drs. Nicholas Ashford (MIT), Theo Colborn, (World Wildlife Fund) and John Wargo (Yale University) presented scientific data which showed that chemical exposures in homes, pollutants in the air, food, soil and water (including pesticides) play a part in causing developmental problems in children.

Presentations on food allergies/sensitivities and ADHD were made by Drs. Sami Bahna (University of South Florida), William Kniker (University of Texas), Michael Jacobson (Center for Science in the Public Interest) and Marvin Boris, pediatrician and allergist (Woodbury, New York.)

Drs. Joseph Bellanti and Aderbal Sabra (Universidade do Grande, Rio School of Medicine, Rio de Janeiro) discussed the gut (intestinal tract) and immunological mechanisms which affect the nervous system. Dr. Randi Hagerman described the physical, behavioral and cognitive manifestations of many syndromes which cause ADHD. Dr. Michael McCann, pediatrician, (Parma, Ohio) presented data which showed that pancreatic enzymes and probiotics may help in treating and preventing food allergies.

Dr. Sidney Baker, a former member of the pediatric faculty, Yale University, discussed the importance of treating ADHD with multiple therapies. He also said, *"Rather than putting a label on a child, it is more important to ask these questions, 'What is the child lacking that he needs?' and 'What is the child receiving that he doesn't need?'"*

Dr. Richard Layton reviewed many therapies which have been proposed for the management of ADHD and described the ten interventions he had found most effective in his practice. Dr. Glen Elliot (University of California) described his research on the relationship of diet and nutritional supplements to ADHD. In my presentation I discussed the relationship of sugar and yeast to ADHD and reviewed a number of reports by professionals with differing points of view.

Participants in this segment of the program included parents, teachers, Dawn Wallerstedt, Family Nurse Practitioner, Karen Duncan, Academic Director of the Lab School of Washington, Kathleen Conroy, a teacher and parent representative (Fishkill, NY), Patricia Frederick, Editor of *Fine Print* (Annapolis, MD) and Michael Severson, President, Minnesota Chapter of American Academy of Pediatrics. Participants in the audience included pediatricians Allan Lieberman (North Charleston, SC) and Linda Rodriguez (Virginia Beach, VA).

The final three-hour segment of the program focused on "Unmet Needs," "Where Do We Go From Here?" and "Recommendations for the Future."

A limited number of the 207-page syllabus of papers presented at the conference are available from IHF, Box 3494, Jackson, TN

38303. A donation of $25 is requested. The conference was recorded by Insta Tapes, P.O. Box 908, Couer d'Alene, ID 83816–0908. Phone: 208–667-0228; Fax: 208–667-6834.

Autism

During the first two decades of my busy pediatric practice, I saw no children with autism or other related developmental disorders. Reports I read in medical journals at the time said, in effect, "Parents are the cause."

Over 30 years ago, Bernard Rimland, Ph.D., the father of an autistic son, published a book on autism. Its purpose was to show that this devastating disorder is *not* caused by the failure of the parents to provide the infant and young child with loving care. Instead, Rimland pointed out that autism develops because of biological disturbances that affect the child's nervous system.

In the 1970s, Rimland established the Autism Research Institute (formerly the Institute for Child Behavior Research), 4182 Adams Ave., San Diego, California 92116. Through this institute he has provided information about the biological causes of autism to people all over the world. He also publishes a newsletter and collects and disseminates information packets and books.

The Yeast Connection to Autism

In the early 1980s I saw a five-year-old boy who had been troubled with recurrent ear infections and hyperactivity during the first two years of life. Yet, his developmental milestones were normal until the age of two and one-half when specialists at a university center made a diagnosis of "pervasive developmental disorder with symptoms of autism."

On a comprehensive treatment program that included nystatin, a special diet and the avoidance of chemical pollutants, the child improved significantly—even dramatically—although he continued to experience developmental problems.

During the 1980s as Rimland and I corresponded, he told me he had received numerous reports from parents whose autistic children improved following anticandida therapy. During the late 1980s, I saw several more children in my practice with autistic-like symptoms who improved on a sugar-free, special diet and nystatin. All through the '90s, I received many, many calls and letters from parents of autistic children. *Almost without exception, autistic symptoms in these children first appeared during the second and third years of life following repeated ear and other infections.*

Autism, like many other chronic and often devastating disorders, develops from many different causes. Yet, there's now clear evidence that in many children, it is yeast connected.

More Support for the Yeast Connection to Autism

The October 1994 issue of *Autism Research Review International,* published by the Autism Research Institute, included a two-page review, "Parent Ratings of the Effectiveness of Drugs and Nutrients." Here are excerpts from this report.

> The parents of autistic children represent a vastly important reservoir of information on the benefits and adverse effects . . . of the large variety of drugs and other interventions that have been tried with their children . . . The data presented in this paper have been collected from the more than 8,700 parents who have filled out questionnaires designed to collect such information . . .
>
> The 31 drugs listed first were prescribed by the child's physician in each case. Note that Ritalin, the drug most often prescribed, is near the bottom of the list. Only 26 percent of the parents reported improvement, while 46 percent said the child got worse on Ritalin.[11]

I studied the graphic charts in the report and I noted that nystatin (or Nizoral) ranked higher than any other prescription drug. Of the 208 children who were given one of these medications, 49 percent found that the child was "better" on the medication and only four percent said that the child was "worse." The better/worse ratio was more than 12 to 1.

By contrast, none of the other drugs showed a better/worse ratio of more than 2.7 to 1 and many prescription medications, including Ritalin, which was given to 1,661 children, showed a better/worse ratio of 0.5 to 1. (This means that twice as many children were made worse by Ritalin than were helped.)

In the five years since Dr. Rimland published his July 1999 questionnaire review, there have been many new developments. Here's a brief summary of some of them:

- In January 1995, the first DEFEAT AUTISM NOW! (DAN!) conference was held in Dallas, Texas. The thirty participants included practicing physicians, parents and researchers from major medical centers in the U.S. and Europe, some of whom were parents or grandparents of autistic children.

  Several speakers described the favorable response of autistic children to treatment programs featuring dietary changes, nutritional supplements (especially Vitamin B_6 and magnesium) and anti-yeast medications.

  One participant, William Shaw, Ph.D., presented clinical and laboratory studies which showed that autism is often yeast-related. His studies were carried out with the collaboration of Enrique Chaves, M.D. and Michael Luxem, Ph.D. These researchers found elevated fungal metabolites and other abnormal organic acids in the urine, especially in children who had taken repeated courses of antibiotic drugs. Following treatment with the antifungal medications, nystatin or Diflucan and dietary changes, all the autistic children improved.[12]
- A 48-page consensus report of the DAN! conference, Biomedical Assessment Options for Children With Autism and Related Problems by Sidney M. Baker, M.D. and Jon Pangborn, Ph.D., was published in January 1996.*
- DAN! conferences for professionals and parents have been held in 1996, 1997, 1998 and 1999. Attendees have included parents and professionals from the United States, Canada, Australia, England and other countries.
- In January 1998, William Shaw published a 300-page book, *Biological Treatment for Autism and PDD,*—a comprehensive

*This report, which was updated in April 1999, discusses many nutritional and biochemical abnormalities other than yeast which play important roles in autism and will be updated periodically. Copies may be obtained for a tax-deductible contribution of $25 or more made to the Autism Research Institute Research Fund.

and easy to read guide to the most current research and medical therapies for autism and PDD. Included in this book are contributions from Dr. Rimland and other professionals and parents of autistic children. Shaw's book can be ordered by calling 913–341-8949.

- A double-blind, placebo-controlled, crossover study is now being carried out by Shaw and a collaborator. Here are several of the objectives of this study:

    1. To determine the incidence of dysbiosis (overgrowth of yeast and abnormal bacteria) in children with autism.
    2. To determine the effectiveness of nystatin in the reduction of autistic behaviors.
    3. To determine if the behavioral improvements resulting from nystatin therapy are maintained over time when nystatin is discontinued.

- In his continuing observations, Shaw has found elevated fungal metabolites and increased numbers of anaerobic bacteria, including *Clostridium difficile,* in many children with ADHD who have received repeated courses of broad spectrum antibiotic drugs.

Why Autism May Be Yeast Connected

There appear to be several mechanisms. One of these appears to be the direct effect of candida toxins on the brain. In a report describing his studies on candida toxin in mice, Iwata, a Japanese mycologist, commented:

> Canditoxins produced unique clinical symptoms. Immediately after . . . intravenous injection (of toxin) animals exhibited ruffled fur and unsettled behavior. Toxicity was so acute and severe that the majority of treated animals succumbed . . . within 48 hours. Within 10 minutes after being given a dose of toxin, the animals became unsettled and irritable; had congestion of the conjunctiva, ears and other parts of the body and finally developed paralysis of the extremities.[13]

In a paper published in another journal describing his research studies on mice, Iwata commented:

> When injected into uninfected mice, Canditoxin exerted toxic manifestation in spleen lymphoid cells . . . This indicates the possibility that . . . the toxin produced in the invaded tissues may act as an immunosuppressant to impair host defense mechanisms involving cellular immunity.[14]

A second way candida may be related to autism is through the disturbance of the normal balance of microorganisms in the intestinal tract. When this occurs, the protective membrane lining the intestines is weakened. As a result, food allergens are absorbed and this may cause adverse reactions in the nervous system.

My Comments

The research studies of Iwata, the clinical data collected by Rimland, and the new research studies by Shaw and colleagues do not prove that the manifestations of all autistic children are yeast-related and/or that antifungal medications provide a "quick fix." Yet, based on reports I've received from both parents and professionals, I feel that all autistic children should be given prescription antifungal medication and a sugar-free diet as an integral part of their management program—especially those children . . .

- Whose developmental status was normal during the first 6–18 months of life.
- Who were treated with broad-spectrum antibiotics for ear or other infections.
- Whose autistic symptoms developed during the second or third year of life.

Rimland's data also showed that nutritional supplements, especially high doses of vitamin B_6 and magnesium, provided significant help to many autistic children.

Breaking the Vicious Cycle of Ear Infections and Preventing the Development of Hyperactivity, Attention Deficits and Autism

Here are my recommendations for breaking the vicious cycle of ear infections and preventing the development of hyperactivity, attention deficits and autism:

1. Don't rush to your physician and ask for an antibiotic when your infant or young child complains of mild ear pain. Instead, use simple home remedies to relieve his discomfort.
2. Change your child's diet . . . eliminate all milk and dairy products . . . and sugar containing beverages and foods. Instead, offer water and a variety of fruit juices.

   You'll find recipes in *The Yeast Connection Cookbook* by William Crook, M.D. and Marjorie Jones, R.N., *Special Diet Solutions* by Carol Fenster, Ph.D. (301-741-5408; www.savorypalate.com) and *Feast Without Yeast* by Bruce Semon, M.D., Ph.D. and Lori Kornblum.
3. When your child develops a respiratory infection, give extra Vitamin C. You can purchase Vitamin C crystals or powder from a health food store. Add 1 teaspoon to 6 ounces of water. Offer to 1 ounce of this mixture every 1 to 3 hours . . . unless it causes stomach ache or loose stools. You may safely sweeten the mixture by adding a few drops of liquid saccharin. Or you can add it to juice.
4. Purchase liquid Zinc sulfate from your health food store or pharmacy . . . when taken along with the Vitamin C, it strengthens the immune system. Give your child one to two milligrams 4 times a day.
5. If your physician examines your child and suggests an antibiotic for an ear or other respiratory problem, ask him/her if your child could get along without the antibiotic.
6. If your child is given an antibiotic, ask your physician to prescribe oral nystatin powder, 2 ounces. Give your child 250,000 units (approximately 1/16 tsp.) by mouth with each dose of the antibiotic. I do not recommend Mycostatin (Squibb) or Nilstat (Lederle) suspension because they are sweetened with sugar.
7. Go to your health food store or pharmacy and ask for probiotics which contain *Lactobacillus acidophilus* (and other friendly bacteria). Give your child one capsule or 1/4 teaspoon of the powder with each dose of the antibiotic. (See also page 129.)
8. After the course of antibiotics has been completed, I recommend continuing the nystatin and probiotics, two or three

times daily for several weeks. Here's why: nystatin discourages the growth of yeast in the intestinal tract and the probiotics replace important friendly bacteria. These products help heal the "leaky gut," lessen the absorption of milk, wheat and other allergens and decrease the chances of your child developing repeated ear problems.

References

1. Truss, C. O., *The Missing Diagnosis,* Second Edition, P.O. Box 26508, Birmingham, AL 35226, 1986; pp. 77–82.

2. Grundfast, K. and Carney, C. J., *Ear Infections in Your Child,* Warner Books, New York, 1987.

3. Hagerman, R. J. and Falkenstein, M. A., "An Association Between Recurrent Otitis Media in Infancy and Later Hyperactivity," *Clinical Pediatrics,* 1987; Vol. 26 No. 5.

4. Crook, W. G., *Journal of Learning Disabilities,* "Can What a Child Eats Make Him Dull, Stupid or Hyperactive?", 1980; 13:53–58.

5. Egger, J., et al, "Controlled Study of Oligoantigenic Treatment in the Hyperkinetic Syndrome," *The Lancet,* 1985; ii:540–545.

6. Egger, J., et al, "Oligoantigenic diet treatment of children with epilepsy and migraine," *J. of Pediat.,* 1989; 714:51–58.

7. Kaplan, B. J., et al, "Dietary Replacement in Pre-School-Aged Hyperactive Boys," *Pediat.,* 1989; 83:7.

8. Carter, C.M., *et al.,* "Effects of a Few Food Diet in Attention Deficit Disorder," *Archives of Diseases in Childhood,* 69:564–568, 1993.

9. Rowe, K., Rowe, K., "Synthetic Food Coloring and Behavior," *Journal of Pediatrics,* 125:691–698, 1994.

10. Boris, M., and Mandel, F. S., "Foods and Additives Are Common Causes of the Attention Deficit Hyperactive Disorder in Children," 1994.

11. Rimland, B., "Parent Rating of the Effectiveness of Drugs and Nutrients," *Autism Research Review International,* Autism Research Institute, 4182 Adams Ave., San Diego, CA 92116, October, 1994.

12. Shaw, W., Chavis, E., and Luxem, M., "Abnormal Urine Organic Acids Associated with Fungal Metabolism in Urine Samples of Children with Autism: Preliminary Results of a Clinical Trial with Antifungal Drugs," the Proceedings Meeting of the Autism Society of America, 1995.

13. Iwata, K., In *Recent Advances in Medical and Veterinary Mycology,* University of Tokyo Press, 1977.

14. Iwata, K. and Yamamoto, Y., "Glycoprotein Toxins Produced by Candida Albicans." Proceedings of the Fourth International Conference on the Mycoses, June, 1977, PAHO Scientific Publication #356.

CHAPTER · NINE

Steps You'll Need to Take to Regain Your Health

■ ■ ■

Steps One & Two: Believe in Yourself—Take Charge

You can get well!! And believing that you can is the first important step. Say to yourself, "I can and will regain my health." The second step is *taking charge*. Say to yourself, "If it's going to be, it's up to me." Read, study and learn. Be responsible. Although you'll need help from kind and caring health professionals, *you* must make the major decisions.

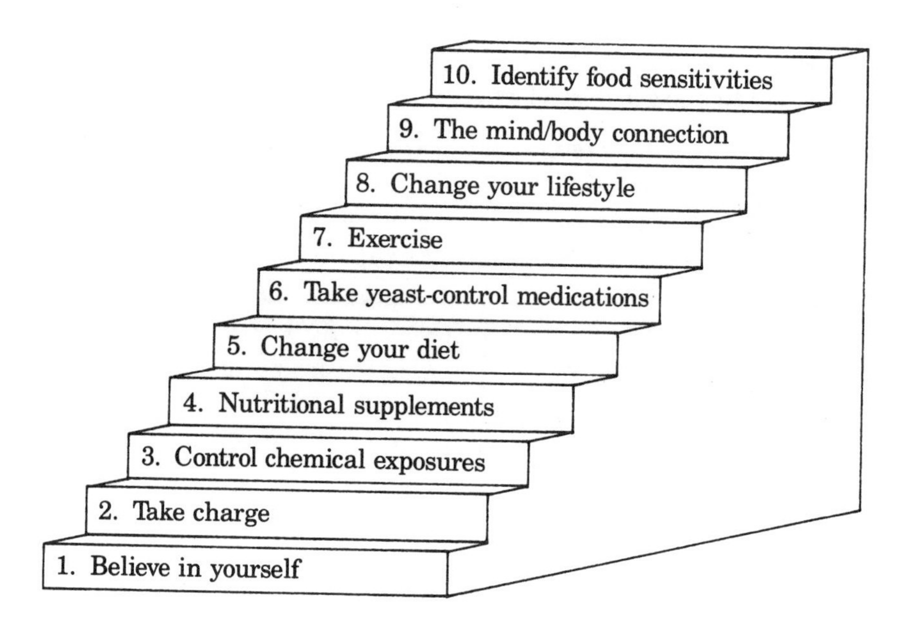

Step Three: Control Chemicals in Your Home

Almost without exception, every person with a yeast-related disorder is made worse by chemicals. So begin at once to get rid of the chemical pollutants in your home. During the past decade, the importance of indoor chemical pollutants has been recognized by a growing number of professionals and nonprofessionals. Here are excerpts from a commentary published in the *Journal of the American Medical Association:*

> The risks Americans may face from some of the pollutants dumped and leaked into the environment by industry may be less than the risk from activities in the home, such as smoking, showering, using room deodorizers and storing and wearing dry-cleaned clothes.[1]

Further recognition of the importance of chemicals in making people sick came in a comprehensive 17-page report in the *Chemical & Engineering News*. In discussing the problem, Bette Hileman said that people with these sensitivities . . .

> . . . usually try to avoid the chemical exposures they believe make them feel sick, such as rooms that have been newly carpeted or recently sprayed with pesticides . . . They may not be able to tolerate the smell of new clothing in a department store or the odors in the aisles of cleaning products in the supermarket and so they avoid those places.[2]

In an article in *Woman's Day,* "Are You Allergic to the Modern World?," Sue Browder[3] described the problems experienced by many people who complained of fatigue, achy muscles, headaches, mental fogginess and other symptoms that are caused by chemical sensitivities. She also cited the observations of Lance Wallace who told her about the new carpet installed in EPA headquarters that made more than 100 people sick. (Their continuing health problems were discussed on the TV program, *60 Minutes,* in 1999.)

What You Can Do to Lessen Your Exposure to Indoor Air Pollutants and Lead a Less Toxic Life

Studies by Dr. William Rea and others show that chemical exposures adversely affect many parts of your immune system.

The more chemicals you're exposed to, the greater your chances of developing health problems. If chemical exposures bother you, Dr. Rea's barrel concept will help you understand what's happening. In explaining his concept to me, he said:

> Chemicals you're exposed to resemble pipes draining into a rain barrel. The barrel represents a resistance. If you continue to be exposed to chemicals, the barrel overflows and you develop symptoms.
>
> Infections of various types, including viral and yeast infections, may also precipitate a "leak" in your barrel, even when the barrel isn't full.[4]

Adapted from William Rea M.D. Used with permission.

Here are things you can do to lessen the chances that chemicals will contribute to your health problems.

- *Don't smoke and don't let other people smoke in your home.* People in homes where others are smoking experience twice as many respiratory infections and other health problems as individuals in smoke-free homes. Moreover, such infections set up a vicious cycle of other health problems. People exposed to second-hand smoke have recently been found to have higher levels of cancer-causing chemicals in their urine.
- *Don't spray insecticides in your home.* You also should store insecticides, paint thinners and other toxic chemicals in the outside shed or garage. Don't keep them under the kitchen sink or in the basement where fumes can leak into the house.
- *Buy all natural fibers,* wool, cotton and silk. If you buy permanent press clothing, or sheets, wash them before using.
- *Air dry-cleaned clothes* and bedspreads and drapes outside before bringing them into your house.
- *Avoid—in every way possible—odorous toxic or potentially toxic substances in your home.* These include paints, formaldehyde, gas cooking stove, and many chemically produced perfumes and colognes.
- *Do 30 to 40 minutes of aerobic exercise a week* followed, if possible, by a sauna. If you feel your body is loaded with toxic chemicals, you may need the help of a professional who can carefully guide this "sweating out" therapy.
- *Get an air purifier* to remove dust, molds and some chemicals.
- *Drink lots of bottled water (in glass)* to help your body excrete chemicals.
- *Have your home tested.* If you suspect that your home is contaminated from pesticides and other chemicals, you can call the Indoor Air Quality Information Clearing House, 1-800-438-4318 for advice on getting help. The EPA also runs a pesticide hotline 1-800-858-7378 and a lead hotline 1-800-LEAD-FYI.
- *Try to bring about changes in your workplace.* If your office is making you sick, read one (or more) of the books listed

below. Bring information to your employer or seek a job in a less polluted environment.

A superb book by Lynn Lawson, *Staying Well in a Toxic World*,[5] is not only authoritative and carefully documented, it is as readable as a paperback novel. In a chapter, "Warning: Your home may be hazardous to your health," Lawson tells the story of a family who became sick when they moved into a new house. She also summarizes the things that people need to do to build a safe home. She then talks about the furnishings in the house and at the end of the chapter she gives nine headings of things to do. She also recommends a number of books.

More Sources of Information

Here are a few of the dozens and dozens of books recommended by Lawson:

- *Your Home, Your Health and Well-Being* by David Rousseau, William J. Rea, M.D., and Jean Enwright, Berkeley, CA, Ten Speed Press, 1989.
- *The Non-Toxic Home and Office: Protecting Yourself and Your Family from Everyday Toxic and Health Hazards* by Debra Lynn Dadd, Los Angeles, Jeremy P. Tarcher, 1992.
- *Success in the Clean Bedroom: A Path to Optimal Health* by Natalie Golos and William J. Rea, M.D., Rochester, NY, Pinnacle Publishers, 1992.
- *Clean and Green: The Complete Guide to Non-Toxic and Environmentally Safe Housekeeping* by Annie Berthold-Bond, Woodstock, NY, Ceres Press, 1990.
- *Non-Toxic, Natural and Earthwise* by Debra Lynn Dadd, Jeremy P. Tarcher, Inc., Los Angeles, CA, 1990.
- *The Cure Is in the Kitchen* by Sherry Rogers, M.D., Prestige Publishers, Box 3161, Syracuse, NY 13220.
- *Chemical Exposures—Low Levels and High Stakes* by Nicholas A. Ashford, Ph.D., and Claudia A. Miller, M.D., Van Nostrand, Reinhold, New York, 1991.

A Final Comment

If you're bothered by chemical sensitivities, rush to your bookstore and buy a copy of *Staying Well in a Toxic World*, published in 1994. If they don't have it in stock, ask them to order it, or call or write the publisher, Lynnword Press, P.O. Box 1732, Evanston, IL 60201 (1-847-866-9630).

Step Four: Nutritional Supplements

Vitamins and Minerals

These nutritional supplements strengthen your immune system. You should begin taking them at once. Increasing numbers of physicians (including some who were once skeptical of their value), are taking them and recommending them to their patients and families. For example, Dr. Art Ulene (who for 15 years appeared on the *Today Show*) made this statement in the Foreword to his book, co-authored by his daughter, Dr. Val Ulene.

> No matter what your age, no matter what your health status, according to new research, optimal doses of vitamins and minerals can improve the state of your health and reduce your chances of developing many diseases and disorders once considered almost unavoidable.[6]

Syndicated columnist Jean Carper,[7] in her carefully researched books tells how vitamins (and other supplements) can combat aging, cancer, heart disease and immune dysfunction. The importance of nutritional supplements in addition to a good diet was noted by internationally known immunologist Robert A. Good, M.D.[8] in the mid 1980s. In the 1990s other professionals with impeccable scientific credentials began to support the use of these supplements, including Dr. Simin Meydani, of the Human Nutrition Research on Aging at Tufts University who commented,

> We used to think about vitamins strictly in terms of what you needed to prevent short-term deficiencies. Now, we're starting to think about what is the optimal level of vitamins for lifelong health and to prevent age-associated diseases.[9]

More Support for Nutritional Supplements in the Late 1990s:

I don't need to tell you that companies producing nutritional supplements are advertising their products several times a day on your TV screen. And scientific support for their value is coming in from many directions. Here are a few examples:

In a December 1998 article from the University of Texas Medical Branch (Galveston), James S. Goodwin, M.D. and Michael R. Tangum, M.D., discussed the attitudes about micronutrient supplements in American medicine. (*Arch. Intern. Med.,* Vol. 158, November 1998, pp. 2187–2191.) Here are excerpts from their article:

> Throughout the 20th Century, American academic medicine has resisted the concept that supplements with micronutrients might have health benefits . . . Part of the resistance stems from the fact that the potential benefits of micronutrients were advanced by outsiders, who took their message directly to the public.

The authors concluded their fascinating article with comments Dr. Goodwin made in an article published over 15 years ago entitled, "The Tomato Effect: Rejection of highly efficacious therapies." (*JAMA,* 1984; 251:2387–2390).

> There are only three important questions when evaluating a potential treatment. Does it work? What are the adverse effects? and How much does it cost? Although more and more physicians are accepting the use of nutritional supplements and other "alternative therapies," there continue to be skeptics.

In his November 1999 newsletter, *Health and Healing,* *Julian Whitaker, M.D., said,

> Numerous studies have shown that diabetics have low cellular levels of magnesium, zinc, vitamin B6, vitamin C and other essential nutrients . . . This is the easiest aspect of diabetic management . . . It simply requires supplementing with high doses of water soluble nutrients to compensate for the drain caused by diabetes.

---

*See page 244 for information about Dr. Whitaker's newsletter.

My Comments

I agree 100 percent with the authorities who emphasize that taking a bunch of vitamin pills and continuing to eat diets loaded with junk foods, sugar and fats will not enable a person to enjoy good health. Yet, I take and recommend vitamin and mineral supplements for members of my family. For my adult patients with yeast-related health problems, I prescribe these daily supplements.

Vitamin A	5,000–10,000 IU
Beta-carotene	15,000 IU
Vitamin B_1	25–100 mgs.
Vitamin B_2	50 mgs.
Niacin	50 mgs.
Niacinamide	100–150 mgs.
Pantothenic acid	100–500 mgs.
Vitamin B_6	25–100 mgs.
Folic acid	200–800 mcgs.
Vitamin B_{12}	100–2000 mcgs.
Biotin	300 mcgs.
Choline (Bitartrate)	100 mgs.
Vitamin C	1000–10,000 mgs.
Vitamin D	100–400 IU
Vitamin E	400–600 IU
Calcium	500 mgs.
Magnesium	500 mgs.
Inositol	100 mgs.
Citrus bioflavonoids	100 mgs.
PABA	50 mgs.
Zinc	15–30 mgs.
Copper	1–2 mgs.
Manganese	20 mgs.
Selenium	100–200 mcgs.
Chromium	200 mcgs.
Molybdenum	100 mcgs.
Vanadium	25 mcgs.
Boron	1 mg.

When supplements are prescribed by a knowledgeable professional, the amounts may vary considerably from those I've outlined. His or her experience, expertise and clinical judgment will override my recommendations. Although the use of vitamin/mineral supplements continues to be "controversial," solid support for their use increases each year.

Essential Fatty Acids (EFAs)

Our bodies are composed of billions of cells of various sizes, shapes and functions. Each cell is surrounded by a membrane composed of special types of fats. We call them *essential fatty acids (EFAs).*

These good fats come directly and only from our foods and they have many diverse functions. Moreover, they are important in preventing health problems of many types, including eczema and other skin disorders, arthritis, heart disease and PMS.

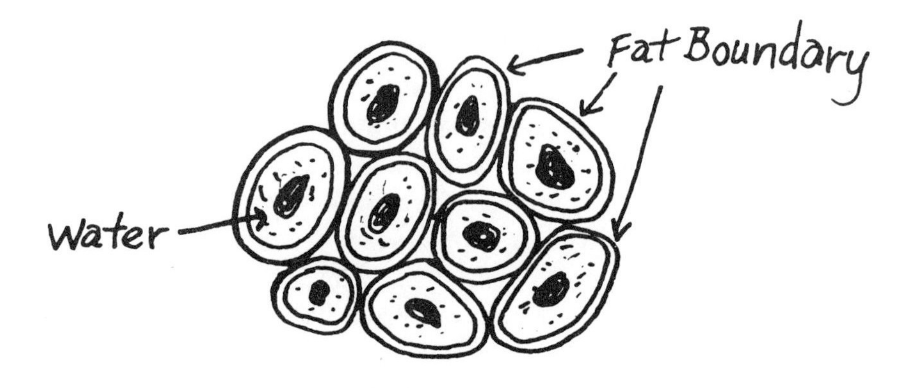

There are two general classes of EFAs—the Omega 3 fatty acids and the Omega 6 fatty acids. Here's how they get their names. Each long chain of fatty acids begins with a carbon atom that has three hydrogens hitched on it. This is called the CH3, or Omega, end of the molecule. Omega 3 fatty acids have the first double bond on the third carbon atom from the CH end of the molecule; the Omega 6 fatty acids have the first double bond on the sixth carbon.

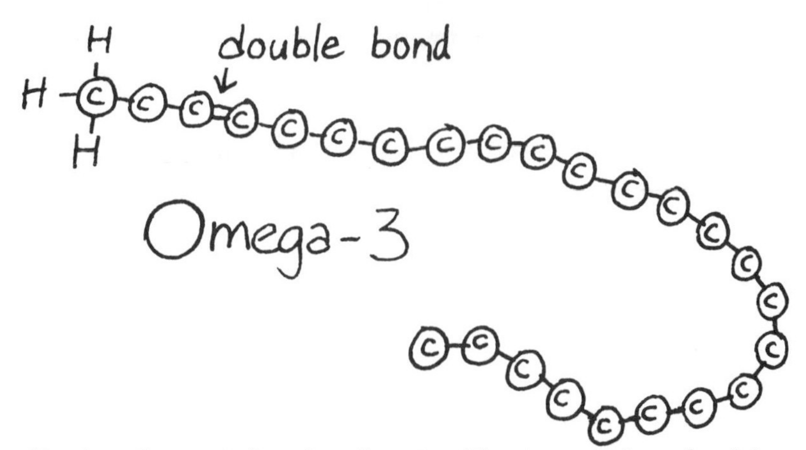

During the past decade, almost without exception, physicians treating patients with yeast-related health problems have used EFA supplements as a part of their treatment program. In the spring of 1994, in responding to my question about the EFA supplements he recommends for his patients, Dr. Leo Galland said:

> The majority of people are getting more than enough Omega 6 oils in their diet. What's lacking in the American diet are Omega 3s. I like to use flax seed oil as a source. Yet, I will use fish oils as well.

The usual dose of flax oil is one to two tablespoons a day. You can mix it with lemon juice and use it as a salad dressing, or you can take it straight. You can get the same essential nutrients from inexpensive organically grown flaxseeds from your health-food store. You must first grind them up in your food grinder and store them in your freezer to keep them from becoming rancid. Then sprinkle a rounded tablespoon on your salad or cereal once a day.

Other Nutritional Supplements

Many other nutritional supplements may help. These include probiotics, CoQ_{10}, *Ginkgo biloba,* garlic, echinacea, bromelain, grape seed extract and other herbal medicines.

For authoritative information on these and other natural remedies:

- Write to the American Botanical Council, P.O. Box 201660, Austin, TX 78720. Fax: 512-231-1924.

- Read Dr. Michael Murray's *Encyclopedia of Nutritional Supplements* and his other wonderful books (see page 247), and an easy-to-read and understand 1999 book, *Vitamins for Dummies* by Christopher Hobbs and Dr. Elson Haas.
- Read nationally syndicated writer Jean Carper's books, *Stop Aging Now* and *Miracle Cures.* These bestsellers are based on cutting-edge research revealing the amazing health benefits of supplements, herbs and foods.

Step Five: Diet

Your next step is *diet.* Compare what you'd need to do if you were taking a three-week ocean voyage on a sailing ship. Before starting your trip you'd have to get enough food. I'm not asking you to buy *all* of it and store it in your pantry and refrigerator, yet here are my suggestions for getting started:

- Go to your kitchen, pantry and refrigerator and get rid of the sugar, corn syrup, white bread and other white flour products, soda pop, most ready to eat cereals, and all the sweet, fat snack foods. *Foods and beverages containing these nutritionally deficient simple carbohydrates encourage yeast overgrowth and promote poor health.* To overcome your candida-related health problems you'll need to avoid them.

- Replace them with more vegetables of all kinds, including some that you may not usually eat. Also go to your health food store and buy the grain alternatives, including amaranth, buckwheat and quinoa. (You'll find instructions and recipes for preparing and serving them in *The Yeast Connection Cookbook.*)

- Get rid of the processed and prepared junk foods, which have hydrogenated or partially hydrogenated fats, as well as those containing food coloring and additives. Add modest amounts of olive, walnut, flax seed, sesame and other unprocessed, unrefined oils.
- Shop mainly around the outer edges of your supermarket. Look for fresh and frozen vegetables, fresh meat, poultry, fish, seafood, eggs, olive oil, pure butter, sardines packed in sardine oil. I especially recommend organically grown foods, which haven't been chemically contaminated. You'll find these foods in many health food stores and in some supermarkets.

What You Can Eat During the First Three Weeks*

Foods You Can Eat Freely

Low carbohydrate vegetables. These vegetables contain lots of fiber and are relatively low in carbohydrates and calories. You can eat them fresh or frozen, cooked or raw.

- Asparagus
- Beet greens
- Broccoli
- Brussels sprouts
- Cabbage
- Carrots
- Cauliflower
- Celery
- Chard, Swiss
- Collard greens
- Cucumbers
- Daikon
- Dandelion
- Eggplant
- Endive
- Garlic
- Green pepper
- Kale
- Kohlrabi
- Leeks
- Lettuce (all varieties)
- Mustard greens
- Okra
- Onions
- Parsley
- Parsnips
- Peppers, bell
- Radishes
- Rutabaga
- Shallot
- Snow Peas
- Soybeans
- Spinach
- String beans
- Tomatoes, fresh
- Turnips

---

*See also pages 150–155 and 158–159.

Meat and eggs

Chicken
Turkey
Beef, lean cuts
Veal
Pork, lean cuts
Lamb
Wild game
Salmon
Mackerel
Cod
Sardines
Tuna
Other fresh or frozen fish
Shrimp, lobster, crab & other seafood

Nuts,* seeds and oils (unprocessed)

Almonds
Brazil nuts
Cashews
Filberts
Flaxseeds
Pecans
Pumpkin seeds
Oils, cold pressed and unrefined
Olive
Safflower
Sunflower
Soy
Walnut
Corn
Butter (in moderation)

Foods You Can Eat Cautiously

High carbohydrate vegetables

Artichoke
Beans, peas and other legumes
Beets
Potatoes, white, baked or boiled
Potatoes, sweet
Winter, acorn or butternut squash
Avocado
Boniata (white sweet potato)
Breadfruit
Celery root (celeriac)
Eggplant
Fennel

Whole grains

Barley
Corn
Kamut
Millet
Oats
Rice
Spelt
Teff
Wheat

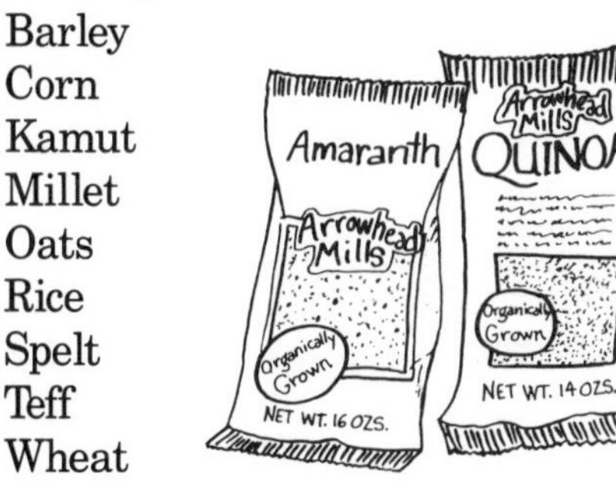

Grain alternatives

Amaranth
Buckwheat
Quinoa

*Since peanuts and pistachios are often contaminated with mold, avoid them.

Breads, biscuits and muffins. All breads, biscuits and muffins should be made with baking powder or baking soda as a leavening agent. You'll find recipes and more information in *The Yeast Connection Cookbook.* Do not use yeast unless you pass the yeast challenge as described on page 106.

Foods You Must Avoid

Sugar and sugar-containing foods. Avoid sugar and other quick-acting carbohydrates, including sucrose, fructose, maltose, lactose, glycogen, glucose, mannitol, sorbitol, galactose, monosaccharides and polysaccharides. Also avoid honey, molasses, maple syrup, maple sugar, date sugar, and turbinado sugar.

Packaged and processed foods. Canned, bottled, boxed and other packaged and processed foods usually contain refined sugar products and other hidden ingredients.

You'll not only need to avoid these sugar-containing foods the early weeks of your diet, *you'll need to avoid them indefinitely.*

Avoid yeast containing foods the first ten days of your diet. Here's a list of foods that contain yeasts or molds:

- **Breads, pastries and other raised bakery goods.**
- **Cheeses:** All cheeses. Moldy cheeses, such as Roquefort, are the worst.
- **Condiments, sauces and vinegar-containing foods:** Mustard, ketchup, Worcestershire®, Accent® (monosodium glutamate); steak, barbecue, chili, shrimp and soy sauces; pickles, pickled vegetables, relishes, green olives, sauerkraut, horseradish, mince meat and tamari. Vinegar and all kinds of vinegar-containing foods, such as mayonnaise and salad dressing. (Freshly squeezed lemon juice may be used as a substitute for vinegar in salad dressings prepared with unprocessed vegetable oil.)
- **Malt products:** Malted milk drinks, cereals and candy. (Malt is sprouted grain that is kiln-dried and used in the preparation of many processed foods and beverages.)
- **Processed and smoked meats:** Pickled and smoked meats and fish, including bacon, ham, sausages, hot dogs, corned beef, pastrami and pickled tongue.

- **Edible fungi:** All types of mushrooms, morels and truffles.
- **Melons:** Watermelon, honeydew melon and, especially, cantaloupe.
- **Dried and candied fruits:** Raisins, apricots, dates, prunes, figs, pineapple.
- **Leftovers:** Molds grow in leftover food unless it's properly refrigerated. Freezing is better.

After you've avoided yeast-containing foods for ten days, you can find out if you're sensitive to yeast by eating a tablet of brewer's yeast, which you can obtain from a health food store.

If it doesn't bother you, eat some moldy cheese.

If consuming these yeasty foods triggers symptoms, stay away from them for several weeks. Then you can experiment further.

Truly yeast-free diets are impossible to come by because you'll find yeast and molds on the surfaces of all fruits, vegetables and grains. Once you've found out that you're sensitive to yeast, you'll need to be your own judge as to how well you tolerate food that may contain some yeasts or molds.

During the first three weeks of your diet avoid fruits. The sugars in fruits, although combined with fiber, are more quickly released and may trigger yeast overgrowth. But, to see if they bother you, you can do the fruit challenge. Here's how:

Take a small bite of banana. Ten minutes later, eat a second bite. If no reaction occurs in the next hour, eat the whole banana.

If you tolerate the banana without developing symptoms, try strawberries, pineapple or apple the next day. If you show no symptoms following these fruit challenges, chances are you can eat fruit in moderation. *But feel your way along and don't overdo it.*

What You Should and Should Not Drink

Water. You should drink eight glasses of water a day. Yet, ordinary tap water may be contaminated with lead, bacteria or parasites. (See pages 175–176 for suggestions for obtaining pure water.)

Fruit juices. These popular beverages are a "no no." Most fruit juices, including frozen, bottled or canned, are prepared from fruits that have been allowed to stand in bins, barrels and other containers for periods ranging from an hour on up to several days or weeks. Although juice processors discard fruits that are obviously spoiled by mold, most fruits used for juice contain some level of mold.

Coffee and tea. These popular beverages, including the health food teas, are prepared from plant products. Although such products are subject to mold contamination, most people seem to tolerate them. To decide, you can experiment. Teas of various kinds, including taheebo (Pau d'Arco) and mathake tea, have been reported to have therapeutic value. If you can't get along without your coffee, limit your intake to one or two cups a day. Drink it plain or sweetened with stevia or liquid saccharin. (See page 182.)

Alcoholic beverages. Wines, beers and other alcoholic beverages contain high levels of yeast contamination, so if you're allergic to yeast, you'll need to avoid them. You should stay away from alcoholic beverages for another reason: They contain large amounts of quick-acting carbohydrate. If you drink such beverages, you'll be feeding your yeast.

Diet drinks. These beverages possess no nutritional value. Moreover, they're usually sweetened with aspartame (Nutrasweet), which causes adverse reactions in many people. They also may contain caffeine, food coloring, phosphates and other ingredients, which disagree with many individuals. However, since diet drinks do not contain mold, some people with candida-related problems may tolerate them. If you drink them, use them sparingly.

Meals

The menus listed on the next few pages are all *sugar-free* and *yeast-free* and designed to help you answer that always troublesome question, *What can my family and I eat?*

The menus for the early weeks are also *fruit-free* and contain relatively few grains and high carbohydrate vegetables (such as potatoes, yams and lima beans). Depending on your likes and dislikes, using these general guidelines, you can change these menus to suit your tastes and those of other members of your family. (You'll also find some simple recipes in Chapter 14 of this book.)

If you pass "the yeast challenge," you can cautiously add cheeses, mushrooms, or other yeast-containing foods to your diet on a rotated basis. *You should also rotate your other foods, espe-*

cially during the early weeks and months of your treatment program. Here's why:

Many, and perhaps most, individuals with yeast-connected health problems are allergic to several (and sometimes many) different foods. The more frequently you eat a particular food, the greater your chances of developing a "hidden" allergy to that food. Such an allergy may contribute to your fatigue, headaches, muscle aches, depression or other symptoms. And strange as it may seem, you may become addicted to the foods that are causing your symptoms, so you crave them.

You'll find a further discussion of food allergies and sensitivities and detailed instructions for carrying out elimination/challenge diets in Step 10 of this chapter. You can find menus and suggestions that will make carrying out your diet detective work much easier in *The Yeast Connection Cookbook,** which I co-authored with Majorie Hurt Jones, R.N. This book contains more than 225 recipes, which will help you with your meal planning.

The meal suggestions that follow, and the recipes in Chapter 14, are a "bare bones" guide to help you get through the initial stages of the diet. In the months ahead, even if you are an inexperienced cook, you'll find that what Marge has to say will help you select and prepare foods that will please you and other members of your family.

Meal Suggestions for the Early Weeks

Breakfasts

- Oatmeal with butter or flaxseed oil and pecans.
- Brown rice with filberts, sardines packed in sardine oil and rice cakes.

*Although I am listed as the co-author of this book, I'm the first to admit, I'm no cook! So I teamed up with Marjorie Hurt Jones, R.N., the author of two superb cookbooks, including the *Allergy Self-Help Cookbook* (Rodale Press) and *Allergy Recipes: Baking with Amaranth.* Marge tested all the recipes in *The Yeast Connection Cookbook.* In a special section she discusses meal planning for the person who isn't improving. Included in this section are menus for rotated diets, additional ideas for breakfast, lunch and supper/dinner, menus for diets featuring various kinds of meats, vegetarian meal suggestions, desserts, afternoon snacks, ingredients and techniques and much, much more.

- Well cooked eggs, 2 strips of crisp bacon and grits with butter.
- Oatmeal, pork chops and cashew nuts.
- Cooked amaranth, tuna, tomatoes and walnuts.
- Cooked quinoa, baked sweet potato and pecans.

Lunches

- Pork chop with broccoli, sesame/oat crackers, sliced potatoes.
- Beef patty, one cup of string beans, filberts and steamed cauliflower.
- Turkey breast, baked sweet potato.
- Salmon, carrots, rice cakes, turnips, cabbage.
- Vegetable soup (or Progresso lentil soup), rice cakes, almonds.
- Tuna fish with lemon and chopped celery; pecans; salad with tomato, lettuce, green pepper, cucumber and radish; flaxseed oil and lemon juice dressing; rice cake.

Suppers

- Baked Cornish hen, steamed cabbage, asparagus, salad with lettuce and pecans. Use walnut oil and lime juice dressing.
- Steak (or hamburger patty), eggplant, mixed green salad with cucumbers and green peppers.
- Pork chops or lamb chops, turnip greens, okra, carrot and celery sticks.
- Roast turkey, baked acorn squash, steamed spinach, grated cabbage, almonds with lemon juice and flax oil dressing.
- Mixed vegetables cooked in microwave, pecans.
- Tuna fish, broccoli, black-eyed peas.

After the first several weeks of your diet, you can experiment. And, chances are you can eat freely . . .

- All fresh vegetables.
- All fresh fruits (in moderation).
- Whole grains (in moderation).

You can continue to consume fish, lean meat, egg, nuts, seeds and oils.

And if you pass the yeast challenge test, you also can include some of the yeast-containing foods.

You must continue to avoid . . .

- Sugar, maple syrup, honey, corn sugar, date sugar and sugar-containing foods: Packaged and processed foods of low nutritional quality that contain sugar and hydrogenated or partially hydrogenated fats and oils.

Meal Suggestions After the First Few Weeks

Breakfasts

- Ground beef patty, scrambled eggs, grits with butter, applesauce muffin.
- Pork chop, steamed brussels sprouts, whole wheat biscuit, grapefruit.
- Toasted rice cakes with peanut butter, sliced banana, turkey burger.
- Brown rice with butter and chopped almonds, tuna (water packed), fresh pineapple.
- Eggs, well cooked, any style; pancakes made with teff, spelt, kamut; freshly squeezed orange juice.
- Barley cereal with banana and pecans, milk, fish (baked or broiled).
- Hot oatmeal with cashews, milk and fresh or frozen peaches.

Lunches

- Salmon patty, corn bread, boiled cabbage, blackeyed peas, sliced tomatoes, orange.
- Fish cakes, steamed cauliflower, boiled okra, rice cakes, strawberries.
- Tuna salad on lettuce, rice cakes, steamed green beans, boiled brussels sprouts, fresh pineapple.
- Swiss steak, steamed artichoke, turnip greens, raw carrots, corn bread.
- Chicken salad, rice soup, spinach, rice biscuits, apple.

- Pork chop, lettuce & tomato salad, applesauce muffin, baked banana.
- Meat loaf, barley soup, celery and carrots, whole wheat biscuits, pear.

Suppers

- Sauteed liver, lima beans, baked acorn squash, sliced tomato, banana oat cake.
- Broiled fish, cabbage & carrot slaw, wax beans, whole wheat popovers, baked banana.
- Broiled lamb chops, steamed cauliflower, steamed broccoli, boiled potatoes, baked apple.
- Rock Cornish hen, steamed carrots and peas, wild rice, rice crackers.
- Roast duck, kale, barley soup, sweet potato, steamed green beans, corn bread.
- Broiled steak; baked potato; lettuce, tomato, cucumber salad with freshly squeezed lemon juice and safflower or linseed oil dressing; mixed greens; fresh strawberries.
- Chicken and rice, steamed artichoke, turnip greens, corn bread, pear.

Shopping Tips

- Feature whole foods.
- Avoid foods labeled "enriched" if you're allergic to yeast.
- Use fresh fruits and vegetables. Commercially canned products often contain yeasts and added sugar. Buy fresh organic vegetables when possible.
- Since many, and perhaps most canned, packaged and processed foods contain hidden ingredients, including sugar, dextrose and other carbohydrate products, avoid them.
- If you must use canned or packaged foods, *read labels carefully.*
- If buying frozen vegetables, select those without added ingredients such as fancy sauces.
- Avoid processed, smoked or cured meats, such as salami, wieners, bacon, sausage and hotdogs, since they often contain

sugar, spices, yeast and other additives. Such foods also are loaded with the wrong kind of fat.

- Avoid bottled, frozen and canned juices. If you want juice, buy fresh fruit and prepare your own.
- Buy nuts from a natural food store where there is a rapid turnover and the nuts are less apt to be rancid or contaminated with molds. Store them in your refrigerator or freezer. Avoid peanuts if you're allergic to yeasts or molds.
- All commercial breads, cakes and crackers contain yeast. If you wish yeast-free breads, you'll have to obtain them from a special bakery or bake your own. Word of caution: Many people with yeast-related problems react adversely to wheat. So, if you continue to experience symptoms, you may need to avoid breads and similar products. Hain and Chico San or Golden Harvest Rice Cakes contain no sugar or yeast. Most rice cakes contain no sugar. These include those produced by Hain and Chico San and Golden Harvest.
- Use expeller-pressed vegetable oils, such as sunflower, safflower, flaxseed and corn. Flaxseed oil is a superb source of the important Omega 3 essential fatty acids. To make salad dressing, combine the oil with fresh lemon juice to taste.
- Buy whole grains (barley, corn, kamut, millet, oats, rice, spelt, teff and wheat) from a natural food store. Grains can be an important ingredient of a nutritious breakfast. Barley, rice and other grains can also be used in various ways at other meals. Barley or rice casseroles are especially tasty.

More Helpful Suggestions

If your health problems are yeast-connected, you may improve . . . often dramatically . . . when you stop eating foods containing significant amounts of cane sugar, beet sugar, corn syrup, fructose, dextrose or honey. Then if you follow other parts of the candida-control program, after two or three months you may find you can consume foods that contain a small amount of sugar.

Yet, if you're allergic to yeast and molds, you may pay for any dietary infraction. And you may not achieve maximum improvement until you avoid all foods that contain yeast and molds. So you may need to stay away from spices, sprouts, condiments and unfrozen leftover foods. (Mold quickly grows on any food that isn't eaten as soon as it's prepared.)

Then if you're still experiencing problems, you'll need to carry out food allergy detective work. In so doing, you identify and avoid all foods that cause adverse or allergic reactions. Common offenders include milk, egg, wheat, corn and soy. However, any food can be a troublemaker. Identifying hidden food allergies requires a carefully designed and appropriately executed elimination diet.* (See Step 10 of this chapter.)

Every person differs from every other person. *You are unique.* In following the anti-candida diet, use a trial and error approach. Most of my patients with candida-related illness, as they improve, can follow a less rigid diet, especially if they're following other measures to regain their health. Included are the use of nutritional supplements, exercise, stress reduction and avoiding exposure to environmental chemicals and mold spores.

Eating Out

If you're like most people, you live "on the run" and eat foods away from home. What's the answer? Do the best you can. And during the early weeks and months of your candida-control program, you may need to do a lot of brown-bagging. And when you eat out, you'll need to make your selections carefully, to avoid foods that trigger your symptoms.

Eggs

Fresh, *well-cooked*† whole eggs offer many ingredients needed to build and maintain strong health. They contain all eight essential amino acids, the building blocks of high quality protein, and are rich in essential fatty acids (EFAs). These two groups of nutrients are "essential" in diet because they aren't made by the body. Eat eggs hard boiled, poached or scrambled. Don't fry them because when you cook them at high temperatures the fat breaks

*Blood tests which may help your physician identify foods which cause sensitivity reactions are now available. (See page 203.)

†Important because eggs may be contaminated with salmonella bacteria.

down and produces harmful substances called trans-fatty acids. (See also pages 99–100.)

Vegetables

These plant foods are good for you. And there are many you can choose, including cabbage, broccoli, collard greens, soy beans, brussel sprouts, onions, carrots, tomatoes and many others. You should eat five (or more) servings *every day.* Here are some of the reasons why vegetables play a critically important role in promoting optimal health:

- They're loaded with vitamins, minerals, fiber and *complex* carbohydrates.
- They contain *phytochemicals.* There are hundreds of these substances in plant foods. You've probably read or heard about some of them, including genistein in soy beans, flavones in dried beans and indoles and isothiocyanates in broccoli. There are many more, some not yet identified.
- They contain adequate protein.

Vegetarian diets lessen your chances of developing osteoporosis, heart disease, diabetes and other degenerative diseases, which affect tens of millions of Americans who consume high protein, high fat diets.

Although I included meats in the menus for the early weeks, I urge you to improve the quality of your diet in the months ahead by eating less meat. Here's one reason: *Animal foods are loaded with pesticide residues.* So eat more plant foods, including vegetables, fruits and whole grains. You'll find a comprehensive discussion of the reasons why such diets promote good health in many other publications. (See pages 244–248.)

Sweetened Foods and Beverages

Here's why they cause symptoms:

- When you eat simple sugars, you encourage yeast overgrowth in your digestive tract. A 1993 study in mice showed that intestinal growth of candida was approximately 200 times greater in mice receiving dextrose than in a control group that didn't receive sugar.

- Diets containing large amounts of refined sugar cause your pancreas to put out extra insulin. As a result, rapid up and down fluctuations occur in your blood and brain sugar levels, producing nervousness, weakness, irritability, drowsiness and other symptoms of hypoglycemia.
- When you fill up on sugar-laden foods, you're apt to take in insufficient amounts of essential nutrients, including vitamins, minerals, essential fatty acids and phytopharmaceuticals (found in fruits and vegetables). Such nutrients participate in various body enzyme systems and serve as precursors in the manufacture of hormones and neurotransmitters (chemicals your brain requires to function properly).
- You may be allergic to sucrose and other sugars derived from a particular botanical source (cane, beet, corn or maple).

Yeast and Mold-containing Foods

Avoiding yeasts and molds in your diet isn't easy. Molds are everywhere . . . indoors and outdoors. Although dampness and darkness promote mold growth, as do basements and cellars, molds can grow on any food, including fruits, vegetables, nuts, meats, spices and leftovers.

Although heating . . . even boiling or processing . . . may kill live molds, mold products may be left behind and this may cause problems for some individuals with candida-related disorders.

Eating a yeast-containing food doesn't make candida organisms multiply. So when you develop symptoms from eating a yeasty food, you develop them because you're allergic to yeast products.

Leftovers: Leftover foods provide a rich breeding ground for yeasts and molds. Molds are one of the major microorganisms causing foods to spoil, and all foods spoil. Although refrigeration retards mold growth, even refrigerated foods develop mold contamination. So prepare only as much food as you need and eat it promptly, or freeze leftovers.

Spices & condiments: These dietary ingredients are usually loaded with mold and should be avoided or approached with caution. Limited quantities of salt and juice from a freshly squeezed lemon are your safest food flavoring agents. And freshly squeezed lemon juice, plus unprocessed vegetable oil, makes a healthy, nutritious salad dressing.

Dry Cereals

These cereals, which you'll find in your supermarket . . . even the best of them . . . have been processed and subjected to high heat. Accordingly, they're much less desirable than hot cereals you prepare at home made from whole grain. Moreover, most of these cereals are loaded with sugar and contain malt and added yeast-derived B vitamins. So if you're allergic to yeast, you'll need to avoid them.

If you like dry cereals, you can find some nutritious ones at a health food store. Many of them are organically produced; some of them contain a mixture of grains; and most are fruit-sweetened. I especially recommend the Health Valley cereals.

If you purchase a dry cereal at your usual grocery store, I suggest sugar-free, yeast-free Shredded Wheat. Other cereals that may be suitable include Cheerios, Puffed Rice, Wheat Chex, Puffed Wheat, Post Toasties, Product 19, and Special K. All have less than 6 percent added sugar. I don't recommend them for the early weeks, but as you improve, you *may* be able to tolerate these cereals in limited amounts.

Diversify or Rotate Your Diet

Even if you aren't able to carry out a carefully designed and executed elimination diet, by rotating your foods you may be able to identify some that may be disagreeing with you and causing your symptoms.

In rotating your diet, you eat a food only once every four to seven days. For example, in rotating fruits, you'd eat oranges on Monday, bananas on Tuesday, apples on Wednesday and pineapple on Thursday. Then on Friday you could start over again with oranges (or a related food such as grapefruit). Do the same thing with other food groups, including meats, vegetables and grains.

You'll find more information about rotation diets in Chapter 13.

Suggestions for Breakfast and Eating on the Run

Breakfast is your most difficult meal, especially during the early weeks. In the recipe section of *The Yeast Connection Cookbook,* my collaborator and co-author, Marge Jones, commented:

> What can I eat for breakfast? Time and time again I hear this and it is a valid concern for people on a diet to control candida.
>
> If you strip breakfast of yeast, you'll eliminate toast, French toast, bagels, sweet rolls, even sourdough bread. If you also omit wheat, milk, corn, sugar, soy and egg, you end up with a gaping void in your menus where breakfast used to be.

Marge then gives you lots of breakfast suggestions, including muffins, pancakes or flat bread that you can make a few days ahead, package and freeze. Then when you're ready, you can put them in your toaster oven.

> Eat these versatile little treasures with your fingers, like dainty pieces of toast, or top them with your favorite filling or bean dip for open-faced mini-sandwiches. They go brown bagging easily and travel well on trips, too.

Marge is a real authority on the non-grain alternatives, amaranth, buckwheat and quinoa. These nutritious foods are high in protein, vitamins, minerals and fiber, as well as rich in complex carbohydrates. They're also useful for people who are sensitive to wheat and corn.

She also tells you how to fix breakfast "pudding." Ingredients in this recipe include foods that you may not consider breakfast foods, such as sweet potatoes and other vegetables. Another suggestion, especially during the diagnostic phase of your diet, is a big bowl of oatmeal with chopped nuts or well-cooked eggs with lean meat.

When eating on the run, plan ahead. Don't wait until you're rushing off to work. Make sure that you can fill your "brown bag" with nutritious foods, including raw vegetables, nuts and rice cakes. You also may get some of the nutritious vegetable-based burgers from your health food store.

If you have trouble planning your menus, Marge gives you many suggestions on pages 333–343 of *The Yeast Connection Cookbook.*

Step Six: Antiyeast Medications

Prescription Antiyeast Medications

If your health problems are yeast related, changing your diet is an important step up the stairs that lead to good health. Prescription antifungal medications* are an equally important step. Four of these medications are available in North America: Nystatin (Mycostatin, Squibb and Nilstat, Lederle); and the azole drugs, fluconazole (Diflucan, Pfizer), itraconazole (Sporanox, Janssen) and ketoconozole (Nizoral, Janssen). A fifth antifungal medication, oral amphotericin B (Fungizone, Squibb), "a close cousin" of nystatin, is widely available in Europe and can be obtained from a few pharmacies in the United States. (See page 126.)

All four of these medications are effective in "knocking out" or limiting the growth of yeast organisms, including *Candida albicans.* As is the case with other prescription drugs, they may rarely cause significant adverse reactions. Yet, based on the observations of physicians who have used them in treating thousands of patients, they are remarkably safe and highly effective.

A sugar-free special diet and nonprescription antiyeast agents may help control yeast overgrowth if your health problems are mild and of short duration. Yet, if you're troubled by a moderate to severe yeast-related disorder, I strongly recommend a therapeutic trial of one or more of the prescription medications.

Which medication (or medications) should you take? How long should you take them? These decisions will, of course, be made by your own physician, and will be based on the duration of your health problems and his/her experiences and preferences. During 1995, 1996, 1997 and 1998 I interviewed many physicians who

*These medications must be prescribed by an M.D. or D.O. However, in some states they may be prescribed by advanced practice nurses (APNs), including nurse practitioners (NPs). See also pages 236–237.

were knowledgeable and experienced in treating patients with candida-related health problems.

All of these physicians had treated hundreds—and some had treated thousands—of patients with candida-related disorders. Some prefer a trial of nystatin for several weeks, then if their patient is improving, they continue it for many months. Yet, if a patient is not improving, they switch to one of the azole drugs.

Other physicians begin with an azole drug. The first choice for most of these physicians is Diflucan. However, some prefer Sporanox, especially in individuals with skin problems and fungal infections of the nail. And a few physicians favor Nizoral because it is less expensive.

Nystatin

According to a 1983 book published by Cornell University Press[10] in the late '40s, researchers from the New York State Department of Health discovered a mold in the soil that "knocked out" other yeasts and molds, including *Candida albicans.* Moreover, tests showed that this substance did not harm animals. The researchers, Rachel Brown and Elizabeth Hazen, named the substance they discovered nystatin (pronounced ny-state-in) in honor of the New York State Department of Health. These unselfish scientists donated their discovery to a nonprofit foundation, which licensed it to E. R. Squibb and Sons.

This pharmaceutical company studied this new fungus-fighting substance, patented it, produced it and made it available to physicians and patients all over the world. Following the expiration of the patent in 1974, nystatin has been produced by other pharmaceutical companies in the United States and other countries. Here's information about nystatin published in the authoritative Physician's Desk Reference (PDR):

> Nystatin is an antifungal antibiotic which is both fungistatic and fungicidal, *in vitro* against a wide variety of yeasts and yeast-like fungi. It is a polyene antibiotic of undetermined structural formula, that is obtained from *Streptomyces noursei* . . . It is the first well-tolerated antifungal antibiotic of dependable efficacy for the treatment of cutaneous, oral and intestinal infections caused by *Candida albicans* and other candida species.

> Following oral administration, nystatin is sparingly absorbed with no detectable blood levels when given in the recommended doses. Most of the orally administered nystatin is passed unchanged in the stool . . .
>
> *Candida albicans* demonstrates no significant resistance to nystatin in vitro on repeated subculture in increasing levels of nystatin . . . *Nystatin is virtually nontoxic and nonsensitizing and is well tolerated by all age groups, including debilitated infants, even on prolonged administration* (emphasis added). Large oral doses have occasionally produced diarrhea, gastrointestinal distress, nausea and vomiting.
>
> *No adverse effects or complications have been attributed to nystatin in infants born to women treated with nystatin*[11] (emphasis added).

Nystatin is generally available in 500,000 unit tablets, and sugar-containing oral suspension and topical powder at pharmacies throughout the United States and Canada. Some physicians prefer the tablets because of their availability and convenience for the patient. Moreover, they feel that because the tablet is slowly dissolved, it may be more effective in controlling yeast in the lower part of the intestine.

Along with most physicians interested in yeast-related disorders, I prefer nystatin oral powder. Here's why:

- Yeasts live in your digestive tract from your mouth to your anus. Accordingly, the powder helps get rid of yeasts in the mouth, esophagus and stomach, as well as in your intestines.
- The powder contains no food coloring, chemicals or other ingredients that may cause adverse reactions in chemically sensitive patients.
- It is more economical.

Here are the instructions I give my patients:

Add the prescribed dose of powder to a half glass of water. Take a large mouthful and swish it around before swallowing. Then, drink the rest of the dose, or dump the powder on the tongue and let it dissolve. Although nystatin is bitter, the taste can be improved by combining it with ¼–½ teaspoon of FOS (frutcooligosaccharides). Besides improving the palatability of the nystatin, this naturally sweet substance encourages the growth of friendly bacteria in your intestinal tract.

Since taking the powder when at work or travelling may be inconvenient, I prescribe tablets (or dye-free capsules) for the two daytime doses and the powder for morning and evening.

I usually use and recommend 1,000,000 units, four times daily. If my patient is improving after a few weeks (or months) I reduce the dose to 500,000 units, three or four times daily. If symptoms are continuing, I may increase the dose to 2,000,000 units, four times a day, and/or add an azole drug. (See also pages 249–250.)

Nystatin powder and dye-free capsules can be obtained from the following pharmacies:

The Apothecary
Bethesda, MD 20814
(800) 869-9159
FAX (301) 493-4671

Bio Tech Pharmacal
Fayetteville, AR 72702
(800) 345-1199
FAX (501) 443-5643

College Pharmacy
Colorado Springs, CO 80903
(800) 888-9358
FAX (800) 556-5893

Freeda Pharmaceuticals
New York, NY 10017
(800) 777-3737
FAX (212) 685-7297

Medical Tower Pharmacy
Birmingham, AL 35205
(205) 933-7381

Hopewell Pharmacy and Compounding
Hopewell, NJ 08525
(800) 792-6670
FAX (800) 417-3864

N.E.E.D.S.
Syracuse, NY 13209
(800) 634-1380
FAX (315) 488-6336

Wellness Health & Pharmaceuticals
Birmingham, AL 35209
(800) 227-2627
FAX (205) 871-2568

Willner Chemist
New York, NY 10017
(800) 633-1106
FAX (212) 682-6192

One-eighth teaspoon of powder = 500,000 units = 1 tablet. The approximate cost for 500,000 units of nystatin powder ranges from 15¢ to 30¢ (in 1996).

A special word to the physician about nystatin powder. If you're prescribing nystatin powder for many of your patients, ask your pharmacists to order it in bulk. They can obtain it from the following sources:

- Paddock Laboratories, 3940 Quebec Ave., No., Minneapolis, MN 55427. Phone: 612-546-4676. Fax: 612-546-4842. Available in quantities of 50 million units, 150 million units, 500 million units, one billion units, 2 billion units, 5 billion units.
- Meridian Pharmaceuticals, Inc., 4015 River Road, Amarillo, Texas 79108. Phone: 1-800-687-7850. Fax: 1-800-687-8902. Available in multiple bulk units.

If you're prescribing nystatin powder for your patients only occasionally and your pharmacist does not wish to stock it, your patients can fill their prescriptions from the pharmacies listed above.

Do not use Nystatin topical powder. For many years nystatin powder has been available in combination with talc and other ingredients. And many physicians have prescribed it for their patients with jock itch, diaper rashes and athlete's foot. Yet, in spite of instructions on the package that say "for topical use only," some pharmacies have filled prescriptions for nystatin powder with the topical nystatin.

Although a pharmacist friend of mine told me that taking the topical powder orally would not be hazardous, it's obviously better to obtain "the real thing."

Why you should not use Nystatin suspension. Nystatin suspension (Nilstat, Lederle and Mycostatin, Squibb) has been available for many years. Moreover, these products are often prescribed by pediatricians for infants with oral thrush. Although they sometimes seem to be effective, I do not recommend them. They're loaded with sugar and recent research studies show that sugar greatly increases the growth of *Candida albicans.*

My Comments

A word about "die-off" reactions. Many people who take nystatin or other prescription or nonprescription antiyeast medications feel worse for several days after they start taking them. Symptoms include fatigue, depression, aching, irritability and abdominal pain. Yet, when they continue the medication, the symptoms disappear. Such symptoms are due to "die-off reactions." Here's how I explain these reactions to my patients:

Suppose someone throws a lighted cigarette into a wastebasket full of trash. When it bursts into flames, you rush for your fire extinguisher. You put out the fire, but smoke is produced, which makes your eyes burn, your nose run and causes you to cough. Although you've put out the fire, it takes a while for the smoke to get out of the room. Similarly, when you kill many yeasts in your digestive tract with nystatin or other antifungal drugs, it may take a few days (or longer) to get rid of the dead yeast products.

Follow a strict diet for a week or more before beginning anti-yeast medication to lessen the possibility of such reactions.

The Systemic "Azole" Anticandida Drugs—Nizoral, Diflucan, Sporanox*

These highly effective anticandida medications are synthetic compounds that are classified as imidazoles (Nizoral) or triazoles (Diflucan and Sporanox), according to whether they contain two or three nitrogen atoms, respectively, in the five-membered azole ring. Their antifungal effects are targeted primarily at ergosterol, the main sterol in the fungal cell membrane.

According to a recent review article in the *New England Journal of Medicine,*[12] these drugs prevent the yeast cells from growing and they inhibit the transformation of the spore forms of yeast to the mycelial form.

Nizoral (ketoconazole)

This potent anticandida medication, available only on prescription, was introduced into the United States in 1981. Using Nizoral in my practice in the 1980s, I was able to help more than 150 of my patients, including many who had not improved on nystatin. And the drug caused no significant adverse effects.

As is the case with any medication, Nizoral may occasionally

*For a discussion of another systemic antifungal drug, see pages 250–251.

cause problems. These include endocrine dysfunction,* elevated liver enzymes, and in rare cases, hepatitis. Accordingly, periodic blood tests to monitor liver function should be carried out on patients who take this drug for one month or longer.

Nizoral (or other azole drugs) may cause dangerous heart problems if taken along with the antihistamine, Seldane.

Diflucan (fluconazole)†

This antifungal medication, which had been used in Europe for many years, was licensed for use in the United States in the spring of 1990. The initial recommendations by the manufacturer focused entirely on treating patients with AIDS, cancer and other disorders characterized by severe immunosuppression. Yet, word of its effectiveness in treating patients with fatigue, depression, headache and other yeast-related problems spread to physicians and patients throughout North America in the early '90s.

In late 1993, the Pfizer pharmaceutical company received an approval letter from the USFDA for use of a single 150 milligram oral Diflucan tablet in the treatment of vaginal candidiasis. And in 1995 and 1996, advertisements have appeared in many medical journals and popular magazines, stating that a single 150 milligram oral Diflucan tablet is well tolerated and is as effective as seven days of intravaginal therapy. Moreover, the advertisements pointed out that the one day treatment with this oral medication was more economical and more convenient to the user.

However, in the fine print discussion of this drug (required by the Federal Drug Administration), the advertisement pointed out that various side effects may occur in people taking Diflucan. These included headache, digestive upsets and abnormalities in liver function tests.

My comments. I was surprised—even amazed—to read the precautions listed in the Diflucan advertisements. Here's why: All of the several dozen candida clinicians that I interviewed told me

*Nizoral may reversibly affect testosterone resulting in loss of libido, impotence, menstrual irregularities and on rare occasions it may cause adrenal insufficiency. See also pages 462–464 of *The Yeast Connection and the Woman.*
†See also pages 159 and 251.

that *side effects from Diflucan were rare, even in patients who took the drug for many weeks and months.*

Because I experienced severe adverse effects to prescription drugs on two occasions, I am cautious in prescribing medications of any type for my patients. I always tell them about side effects and I ask them to call me at once if they feel they're experiencing an adverse reaction. In deciding whether to prescribe any drug, I consider the possible risks, as well as the benefits.

In my patients with serious or long-lasting health problems, including chronic fatigue, depression, persistent headaches and asthma, I do not hesitate to prescribe or recommend a several weeks' course of Diflucan. And in people with more serious disorders, such as multiple sclerosis, I may give it for several months.* (See also pages 157–160.)

Sporanox (itraconazole)

This systemic azole drug is a cousin of Nizoral and an even closer cousin of Diflucan in that both are "triazoles." As is the case with Diflucan, Sporanox (itraconazole) was extensively studied before its release in the United States in the spring of 1993.

Here are excerpts from a 1989 article.

> Itraconazole was shown to be extremely effective in a wide range of superficial and more serious "deep" fungal infections when administered once or twice daily. Generally, greater than 80% of patients with superficial dermatophyte or yeast infections are cured by itraconazole . . . Preliminary findings also indicate that itraconazole may hold promise for the prophylaxis of opportunistic fungal infections . . . in women with chronic recurrent vaginal candidiasis.[13]

In another report, the authors commented:

> Itraconazole, appears to be an even more effective agent with a broader spectrum of action than ketoconazole.

---

*R. Scott Heath, M.D., a Cincinnati neurologist, has found that 200 milligrams of Diflucan daily for four months helped most of his MS patients and caused no significant adverse effects. Phillip Mosbough, M.D., an Indiana urologist used a similar dose in his study of women with interstitial cystitis. See also pages 48–49 and 54–55.

> During the past five years itraconazole has been distributed throughout 48 countries. Cure rates in patients with candidiasis were greater than 80%.
>
> Of the 15,000 patients who have been treated with itraconazole, only 7 percent experienced side effects; gastrointestinal complaints and headaches were the most frequently reported of these.[14]

During late 1995 and early 1996, a number of two and three-page advertisements appeared in both medical journals and popular magazines describing the effectiveness of Sporanox in treating fungal infections of the nails. Yet, as it is with Diflucan, the FDA does not allow the company that produces Sporanox to advertise it to physicians or the public for use in treating people with other yeast-related disorders.

Amphotericin B (Fungizone, Squibb)

This antifungal drug was first isolated by the Squibb Institute for Medical Research many years ago. According to comments in the 1986 edition of the *Physicians' Desk Reference,* amphotericin B is "substantially more active in vitro against candida strains than nystatin. Given orally, it is extremely well tolerated and is virtually nontoxic in prophylactic doses."

Like nystatin, it is poorly absorbed from the gut and has a high degree of activity against candida in the intestinal tract. It also has been widely used by the intravenous route in the treatment of deep-seated fungal infections.

Amphotericin B is readily available in pharmacies in Europe and is favored by several British physicians that I consulted. It is available from only a few pharmacies in the United States, including Wellness Health and Pharmaceuticals (1-800-227-2627) and College Pharmacy (1-800-855-9538).

My comments. To gain more information about their use of prescription antifungal drugs (including the safety and effectiveness of the azole drugs), I interviewed two dozen knowledgeable and experienced practicing physicians from 1995–1999. Each of these physicians had prescribed these drugs for hundreds—even thousands—of their patients. Here's a summary of what I learned.

- These drugs are remarkably safe—much safer than many drugs being prescribed by physicians today for infections, arthritis, depression, hyperactivity and many other disorders.
- Drug preferences: Diflucan was the first choice with most physicians, and Sporanox a close second. However, a number of candida clinicians continue to use Nizoral in treating many of their patients, and several have found it as effective in those who did not respond to Diflucan or Sporanox. A possible advantage of Nizoral: When compared to the other azole drugs, it is much more economical.

  One of my physician consultants, Charles Resseger, D.O., Norwalk, Ohio, commented:

  > My first line of defense is still Diflucan. It's extremely safe. In treating thousands of patients, I've seen no problems. Occasionally, after four or five months of treatment I've seen some people's hair temporarily fall out, which is reversible. When I get a resistant case, I add Sporanox to my treatment program. In some patients who continue to have problems, I add a nonprescription agent, such as caprylic acid or Tricycline.*

- Duration of therapy: It ranged from one to two weeks to several months of daily medication. One experienced physician, Sidney M. Baker, M.D., Weston, Connecticut, said:

  > I customarily use Diflucan for a three-week diagnostic trial. In most patients, I give it for no more than 6 to 10 weeks. Yet, I have a couple of patients who have been on this medication for two years. One little boy whose behavior and school performance is more than satisfactory when he's on Diflucan shows a deterioration in his behavior when the medication is stopped.
  >
  > I've tried to get around keeping him on the Diflucan and have been monitoring his liver function very closely. I figure that having this boy's sanity intact is worth the price.

- Most physicians recommend a trial of diet before using nystatin or an azole drug. However, some physicians prescribe

*This potent formulation of plant concentrates is available from several pharmacies. See page 121.

the medication along with the diet, especially in their patients with severe yeast-related disorders.

- Many physicians prefer nystatin and prescribe it for most of their patients. And some find it so effective that an azole drug is rarely necessary. Other physicians begin with a therapeutic trial of an azole drug (usually Diflucan) for two to four weeks. Then, as the patient improves, nystatin is added to the treatment regimen. If the patient continues to do well, the Diflucan is often discontinued or given less frequently (such as one tablet each week).
- Combination therapy: Elmer Cranton, M.D., Yelm, Washington, recommends nystatin and an azole drug concurrently for all of his yeast patients because he feels that the combination is much more effective.* So do other physicians. (See my comments in Chapter 10.)

Nonprescription Antiyeast Medications

Ideally, your own family physician, internist, gynecologist, dermatologist, or other specialist, will help you and work with you. And, even if your physician is skeptical of "the yeast connection," perhaps he or she will prescribe nystatin or one of the azole drugs, Diflucan, Sporanox or Nizoral.

These antiyeast medications literally punch holes in candida cell membranes. They knock them out and keep them from multiplying and raising large families. Moreover, the azole drugs may also keep the little round yeast cells from putting out branches (mycelia) which can burrow beneath your mucous membranes.

But suppose you do not have a knowledgeable medical doctor (M.D.) or osteopathic physician (D.O.), what then? You may be able to find help from other licensed health professionals, including N.D.s, D.C.s and R.N.s. Fortunately, there are also nonprescription substances and products which help limit or retard the growth of candida in the digestive tract.

*You'll find a comprehensive discussion of prescription antiyeast medications, including the comments and experiences of candida clinicians on pages 451–483 of *The Yeast Connection and the Woman.*

Probiotics

Probiotics are a group of friendly bacteria that help us stay well. They include *Lactobacillus acidophilus* and *Bifidum bacterium*. These bacteria, which are found in yogurt, were first identified about 100 years ago.

During the '80s and early '90s, many professionals I've talked to tell me that they use preparations of various probiotics to help control candida in their digestive tract. Although they are not potent agents for "knocking out yeast," recent reports in a major medical journal show they help. (See pages 11, 100, 248.)

In discussing probiotics, Jeff Bland, Ph.D., told me:

> I think we all recognize that the most important criteria for activity of probiotic substances is that there must be an adequate number of live organisms, they must be reasonably resistant to oxgall (bile) and they must be able to adhere to the gastrointestinal epithelium. If any one of these three criteria is not met, then the activity of the product is limited.

You can find many probiotic products in your health food store. I recommend them as a nutritional supplement for all my patients, especially those with yeast-related health problems. When you purchase them make sure they are stored in a refrigerator.

Caprylic Acid

This substance, a natural occurring fatty acid, was first studied some 40 years ago by Dr. Irene Neuhauser of the University of Illinois, who found that it had antifungal activity. Since it is readily absorbed, it is necessary to take time-released or enteric-coated formulas to allow for a gradual release throughout the entire intestinal tract. Because caprylic acid is a food product, it's available without a prescription.

During the past decade, a number of health professionals have found that caprylic acid products are effective in controlling candida in the intestinal tract. Moreover, some of the professionals I've talked to say, "It's just as good as nystatin."

In February 1996, I received information about a treatment program that combines caprylic acid with psyllium, bentonite and probiotics.* In an article in the July 1995 issue of *The Townsend Letter for Doctors and Patients,* S. Colet Lahoz, R.N., M.S.L.A.C., of White Bear Lake, Minnesota, described her successful use of this product in treating a number of her patients. I obtained more information about this treatment program from a Canadian naturopathic physician, Lois Hare, who commented:

> I found this four-pronged treatment program quite effective. The bentonite acts sort of like a sponge in drawing toxins and other waste products out of the intestine and the psyllium helps clean out any fecal buildup. These products together, along with the probiotics and the caprylic acid, seem to help my yeast patients as one part of a comprehensive treatment program.
>
> Yet a few of my immunocompromised patients may also need a systemic agent such as Diflucan. And in such patients, I refer them to a knowledgeable medical doctor for further management.

A word of caution. Although caprylic acid products discourage the growth of *Candida albicans* in the digestive tract, they are no "cure all" and their use should be supervised by a licensed health professional. Attention also should be paid to other parts of the treatment program, including nutritional deficiencies, endocrine problems, and food and chemical sensitivities.

Citrus Seed Extract

Like nystatin and caprylic acid, this antifungal agent discourages the growth of *Candida albicans* in the intestinal tract. I first heard about the efficacy of citrus seed extract in controlling candida from Drs. Leo Galland and Charles Resseger. Both said that citrus seed extract is as effective as nystatin and caprylic acid and other nonabsorbed antifungal agents in treating patients with the Candida Related Complex. It is also effective against giardiasis and some of the other intestinal parasites.

A number of health professionals I've talked to recommend Tricycline, a product that combines grapefruit seed extract and the herbs artemesia and berberine.

---

*Information about this product can be obtained from Monica O'Kane, Acu-Trol, Inc., 2 Willow Rd., St. Paul, MN 55127; (651) 484-2811; 800-594-4675.

Garlic

Garlic has been widely used for medicinal purposes for centuries. For example, Virgil and Hippocrates mention it as a remedy for pneumonia and snake bite. In looking through the Index Medicos, I found numerous articles from American and foreign literature describing the inhibitory action of garlic on candida organisms.

In the early 1990s, a number of the participants at the First World Congress on the Significance of Garlic and Garlic Constituents presented scientific reports, which documented the effectiveness of garlic as a useful anticandida agent. They pointed out that chopping or crushing garlic triggers the release of medicinally active compounds. Moreover, such compounds appear less likely to cause adverse reactions. They also noted that deodorized garlic products possess health benefits comparable to whole garlic.

More recently, Dr. Benjamin Lau of Loma Linda University, Loma Linda, California, reported the effectiveness of aged garlic extract against *Candida albicans* infections in mice. According to Dr. Lau, this study suggested that the garlic extract strengthens the immune system by helping the body's white blood cells gobble up enemy germs.

As you probably know, many different garlic products, including odor free garlic, are available from health food stores. Like the various automobile makers, each company describes the features of their garlic products. Which preparations are best? I don't know. Yet, I've been impressed by reports that document the efficacy of aged garlic extract (Kyolic).

Tanalbit

This internal intestinal antiseptic consists of natural tannins, combined with zinc. It has been found to be effective in the management of yeast overgrowth in the intestine, acute and chronic diarrhea, colitis, constipation and spastic colon.

In discussing Tanalbit, Murray R. Susser, M.D., Santa Monica, California, said:

> I've been using Tanalbit on candidiasis for about eight years. Of course, I have used many other treatments, including ny-

> statin, caprylic acid and Diflucan. I find that Tanalbit is a very consistent product. It gives good therapeutic results with little side effects.

You can purchase Tanalbit from some health food stores and pharmacies, or you can order it from Scientific Consulting Services, 466 Whitney St., San Leandro, CA 94577 (650-632-2370).

ParaCan

In February 2000 I received information about this natural product which incorporates a blend of herbs that have been used for decades to eliminate microbes and worms. Its main active ingredients include black walnut hulls, wormwood, pumpkin seed, Pau d'Arco, echinacea, barberry, gentian, garlic, olive leaf, cloves, chamomille and thyme.

I've also received reports from other sources that several of the herbs contained in this product help control the overgrowth of *Candida albicans* and other yeasts in the digestive tract. According to the webpage, www.genhealth.com, "ParaCan . . . is an effective and . . . safe way to achieve optimal internal health. However on some occasions where parasitic activity is high, some individuals may experience nausea, weakness or fatigue during the cleansing process."

Although this product is not now available in health food stores in the U.S., it can be ordered from Genesis Health Marketing in Australia. Fax: 612–9663-5310; e-mail: sales@genhealth.com.

Goldenseal

This perennial herb was used by native Americans for a wide variety of conditions, including fighting infections. Its activity has largely been attributed to its high content of biological active alkaloids, berberine, hydrastine and candidine.

According to Michael T. Murray, N.D., in his book, *Natural Alternatives to Over-the-Counter and Prescription Drugs,*[15] the antibiotic activities of goldenseal's alkaloids are well documented in the scientific literature. It has been found to be effective against a variety of organisms, including *Candida albicans.*

Goldenseal is available in powder and extract and in combination with other ingredients.

Aloe Vera

This perennial plant has magenta flowers. Its spike-like leaves are filled with a gel that contains many compounds that possess biological activity. The most important of these appear to be the anthraquinones polysaccharides and prostoglandins.

In another of his books, *The Healing Power of Herbs,*[16] Dr. Murray said that aloe has demonstrated activity against many common bacteria and fungi, including *Candida albicans.* He also cited the observations of Jeffrey Bland, Ph.D., and said,

> Six of ten subjects (who were studied) showed marked alterations in stool cultures after the week-long study. This implies that aloe vera juice may exert some antibacterial or anticandida activity. In the four subjects with positive cultures for *Candida albicans,* there was a reduction in the number of yeast colonies.

Colloidal Silver

During 1990's, I received reports about the value of this substance from a number of different sources. These included phone calls and letters. During a visit to Indianapolis a health food store owner told me that silver products are helping many people with yeast-related problems who came to her store.

Miklos Boczko, M.D., of White Plains, New York, told me he had been cautiously using colloidal solutions of silver in treating some of his yeast patients and he felt that this therapy had helped.

About the same time, Dr. John Parks Trowbridge sent me a 300-page book, *Coll/Ag-30 The Silver Micro Bullet.*[17] In skimming through the book I learned that: solutions of silver have been used as a germicide since the early 1900s; the substance is said to be antibacterial, antiviral, antiparasitic and antifungal. It is available over-the-counter in various dilutions. Yet, before recommending it, I would need more information. Certainly its use should be supervised by a knowledgeable and experienced health professional.

Other Products

A number of other products have been used by people seeking to control candida in their digestive tract. These include *Yeast Fight-*

ers by Twin Lab, which contains probiotics, garlic, caprylic acid and *Candida Cleanse,* by Rainbow Light Nutritional Systems. This product contains Pau d'Arco, acidophilus, herbs and other ingredients.

Do these remedies help? I don't really know. I presume they're safe since they're being advertised and distributed in compliance with current laws. Yet, as is the case with the other nonprescription remedies mentioned in this section, I feel their use should be supervised by a knowledgeable and experienced health professional.

Step Seven: Exercise

Whether or not your health problems are yeast connected, you will need to exercise if you want to enjoy good health. All three of my daughters walk almost every day. One of them, Nancy said, "If I don't get my exercise, I can tell it. I do not feel good."

Helps You Overcome Depression and Fatigue

According to Charles W. Smith, M.D., it is especially important for patients with depression and chronic fatigue to exercise. He especially recommended aerobic exercises, which require the use of oxygen to produce energy. Aerobic exercises he recommended include walking, running, aerobic dance, swimming, cross-country skiing, bicycling and rowing.[18]

In commenting on the immediate benefits of exercise, gynecologist Joe S. McIlhaney, Jr., said:

> There are several direct rewards for anyone who starts a regular program of exercise. FEELING BETTER. When patients complain to me of feeling tired and not having much energy, I assure them if they will begin exercising, within two weeks they will feel like different people.[19]

He points out that exercise also helps people increase their mental alertness and control their weight, and that studies have shown that people who exercise develop fewer illnesses and are less apt to have accidents.

Helps Strengthen Your Bones

In his 1994 book, *Preventing and Reversing Osteoporosis,* Alan Gaby, M.D., Baltimore, said:

> It has long been known that bones develop in a way that resist the forces acted upon them. This means that repeated application of a physical stress to a bone will actually cause that bone to remodel and become stronger.

Gaby cited an international study of hip fractures in women from different countries, which showed that physical activity, or lack of it, played a key role in determining whether a fracture would occur.

He pointed out that although all forms of exercise promote better health, weight bearing exercises, such as walking, running or jumping on a trampoline are the best ways to strengthen your bones. And he said,

> When my sedentary patients begin an exercise program, they almost always begin to feel better in many ways within a month or two.[20]

Helps Prevent Cancer

According to Leslie Burnstein, women who participated in one to three hours of exercise a week over their reproductive lifetime had a 20 percent reduction in breast cancer. And those who participated in more activity each week showed a 60 percent reduction in breast cancer.

What is the mechanism? According to Burnstein, exercise affects hormones that are important in determining a woman's breast cancer risk. Although it's best to begin regular exercise early in life it can provide health benefits to people who begin exercising at any age.[21]

Helps Prevent Heart Disease

The role of exercise in preventing heart disease was stressed by the late Nathan Pritikin more than 20 years ago. During a visit to Pritikin's Longevity Research Institute in California in the late 1970s, I talked to a number of people who had been troubled by severe heart problems, diabetes and other chronic diseases who regained their health and their life on a program that featured walking and dietary changes. More recently, Dean Ornish has confirmed Pritikin's observations.[22]

An article in the April 1994 *Mayo Clinic Healthletter,* cited an August 1993 report in the *Journal of American College of Cardiology,* which described some of the beneficial effects of exercise. Adults who exercised vigorously five or six times a week for a year partially reversed the buildup of cholesterol and fatty acids in their arterial walls.

Helps Women's Reproductive Organs

In discussing exercise in *The New Our Bodies, Ourselves,* Janet Jones pointed out that until recently women were taught that exercise could damage their "internal organs." But today, most people understand that exercise helps women, including those with PMS and during and after pregnancy.[23]

Susan M. Lark, M.D., in her book, *PMS—Self-Help Book,* said that she had learned about the personal benefits of exercise, as well as the help it had given many of her patients. The benefits she cited included lessening of fluid retention, prevention of back pain, improved posture, and relief of anxiety and irritability and other psychological effects.[24]

So there you have it. If you want to look good, feel good and overcome health problems of any sort, you'll need to exercise regularly.

Step Eight: Lifestyle Changes

If your health problems are yeast-related, it undoubtedly will help you to change your diet, get rid of chemical pollutants in your home, and take antiyeast medications and nutritional supplements. Yet, there are other things that you can do, too. In his comprehensive, easy-to-read-and-understand book, *Staying Healthy with Nutrition,*[25] Elson M. Haas, M.D., lists 88 survival suggestions, including these lifestyle changes:

- **Create** a good exercise program. This includes regular exercise, stretching and at least 30 minutes of vigorous activity several times a week.

- **Avoid** excessive sun exposure. With the depletion of the ozone layer and the effect of ultraviolet light, the risks outweigh the benefits.
- **Practice** some sort of stress reduction daily. Meditate, lie down without sleeping, or just sit with eyes closed, breathe deeply, relax for at least 15 to 20 minutes.
- **Reduce** or avoid alcohol use. Alcohol depresses the senses and reduces immune resistance. In addition, chemicals are used in processing most alcohol products.
- **Avoid** habitual drug use, such as consumption of caffeine in coffee, tea or colas, and regular sugar use.
- **Drink** more clean water and less soda, coffee, juice and alcoholic beverages.
- **Wear** more natural-fiber clothes (cotton, rayon and silk), especially if you're sensitive to synthetic materials.
- **Buy** and use organic foods, those that are grown without chemicals, fertilizers and pesticides.
- **Minimize** soft drink use. Substitute water or a combination of fruit juice and carbonated mineral water.
- **Take** antioxidant nutrients.
- **Eat** more cruciferous vegetables and rotate foods to avoid allergic/sensitivity reactions.
- **Obtain** natural lighting or full spectrum lights at work.
- **Take** regular breaks from a computer—walk and stretch, drink water and get fresh air.
- **Learn** some healing arts, such as massage, herbal therapy and nutrition for yourself and others.

Step Nine: The Mind/Body Connection

Although I was aware that psychological factors could contribute to the health problems in some of my patients, I didn't really pay much attention to them until about 10 years ago.

At that time, during a trip to California, I had dinner with the late Doctor Phyllis Saifer, a leader in studying patients with food and environmental sensitivities and yeast-related health problems. Phyllis talked about many things, including the role psychological factors played in many of her patients. She said, in effect:

> Although food and chemical sensitivities and candida play an important role in making people sick, I've found in my own practice that many people with these problems—especially women—give a history of having been physically or sexually abused. And such abuse weakens the immune system and makes them more apt to develop other health problems.

About the same time, I read and re-read Norman Cousins' article published in the *New England Journal of Medicine*[26] and his books, *Anatomy of An Illness*[27] and *The Healing Heart.*[28] In the mid-1980s, I visited Mr. Cousins in his UCLA office twice. He told me about his work with groups of people with arthritis, cancer and other chronic disabling illnesses.

> I always tell them jokes and after about the third joke I have them laughing so hard they have to hold their sides. Then when my session with them is completed, I ask, "How many of you are still hurting as much as you were when you walked into this room?" and no hands are raised.

During the past decade in talking to my patients and the people who write and call me I'd say,

> You need caring, empathetic people to encourage you, work with you and help you, including family members and profession-

> als. Support groups consisting of people who are experiencing similar problems can also help.
>
> You need to be noticed, praised and encouraged. You need smiles, touching, holding, patting and petting. Physical contact stimulates the release of endorphins, a chemical that lessens anxiety and pain.

At a meeting of the American Holistic Medical Association in Kansas City, I heard many fascinating discussions of the mind/-body connection, including presentations by Drs. Dossey, Deepak Chopra and James Gordon.*

When I came home from the meeting, I began reading anything and everything I could get my hands on that had to do with the relationship of psychological stimuli to physical symptoms. Books that I found especially fascinating included *Meaning and Medicine*[29] by Dr. Larry Dossey and *Healing and the Mind*[30] by Bill Moyers and *Mind/Body Medicine—How to Use Your Mind for Better Health* by Daniel Goleman and Joe Gurin.

These books were loaded with fascinating information. Included in the latter book were chapters by many different professionals from major American medical centers. I especially liked a chapter (in the Goleman and Gurin book) by Kenneth R. Pelletier, Ph.D., of the Stanford University School of Medicine, who, in talking about prevention and treatment, said:

> Meditation, visualization, hypnosis, biofeedback and relaxation techniques may help prevent and treat a variety of illnesses.[31]

Still another recent book, *Women's Bodies, Women's Wisdom*[32] by Dr. Christiane Northrup, a former president of the American Holistic Medical Association, provides professional and nonprofessionals with a tremendous amount of information.

Included in her book is a discussion of how she and her colleagues at their Holistic Health Center use their dramatic discoveries in mind/body medicine to show women how to heal by listening to their own bodies' wisdom.

---

*In June 1998, Dr. James Gordon and The Center for Mind/Body Medicine hosted an international conference on Comprehensive Cancer Care Integrating Complementary and Alternative Therapies.

In the chapter, "Steps for Healing," the author lists 12 steps, including:

- Respect and release your emotions; learn to listen to your body.
- Gather information.
- Forgive.
- Acknowledge a higher power or inner wisdom.

In discussing this latter step, Northrup said that our bodies are permeated and nourished by spiritual energy and guidance. She also said that having faith and trust in this reality is an important part of creating health. When a woman has faith in something greater than her intellect or her present circumstances, she gets in touch with her inner source of power.

Another professional, Jessica Rochester, Ph.D., a good friend and a member of the Advisory Board of the International Health Foundation, stressed the importance of the mind/body connection:

> There's a wonderful emerging paradigm in medicine with evidence of this in two areas. The first is increased awareness of the importance of nutrition, lifestyle, exercise and relaxation in relationship to health. The second is the considerable data available demonstrating the connection between the consciousness (the mind or soul or spirit) and the physical part of us, the body.
>
> I try to help people develop their inner strengths so that they can look at the patterns that brought them to where they are in their lives now. Also where they would like to go from here . . . Everything is held in the cells of our body, which are made out of what we eat, drink and breathe. But beyond that, every memory, every experience we have is held in our body.[33]

My Comments

You will need to take many steps to regain your health. They include changing your diet, taking anticandida medication, controlling chemical exposures, making lifestyle changes, taking nutritional supplements, tracking down hidden food sensitivities, obtaining psychological support, and working with a kind, caring and knowledgeable health professional.

Doing all the things you need to do won't be easy, but you can get started today. Remember Confucius' proverb, "A journey of a thousand miles starts with one step."

Step Ten: Food Allergies and Sensitivities

Unusual reactions to substances in a person's diet or environment have been recognized for thousands of years. Yet, it wasn't until 1906 that the term "allergy" was coined by the Austrian pediatrician Clemens von Pirquet.[34] He put together two Greek words, *allos*—meaning "other" and *ergon*—meaning "action." To von Pirquet *allergy* meant altered reactivity.

Some doctors feel the term "allergy" should be limited to those conditions in which an immunological mechanism can be demonstrated using allergy skin tests or more sophisticated laboratory tests. But other conscientious physicians feel that the allergic and hypersensitivity diseases are much broader in scope. These included Frederic Speer, M.D.,[35] University of Kansas, John Gerrard, M.D.,[36] University of Saskatchewan, and William C. Deamer, M.D., University of California, San Francisco.

In his award-winning presentation delivered before the Section on Allergy of the American Academy of Pediatrics, Dr. Deamer commented on the deceptive nature of food allergy and the difficulties in diagnosing and treating it.

Deamer noted the unreliability of food skin tests and the value of trial elimination diets in studying patients with fatigue, irritability, headache, abdominal pain, musculoskeletal discomfort, and respiratory symptoms, including asthma and allergic rhinitis. And, he said:

> There can be no doubt . . . of the role specific foods may play in causing these symptoms nor of the fact that the respiratory tract symptoms may be identical to those caused by accepted antigens.[37]

During the past decade, another university allergist, William T. Kniker, of the University of Texas (San Antonio), has talked and written about the importance of food allergies and sensitivities. In an award-winning lecture and an article, which was published in the *Annals of Allergy,* Dr. Kniker said:

> There are countless—millions—of individuals who have unrecognized adverse reactions to various antigens, foods, chemicals, and environmental or occupational triggers . . . The acquired disease may be limited to body surfaces, or may involve a puzzling array of organ systems causing the patient to visit a number of different specialists who are unsuccessful in recognizing that an allergic or adverse reaction is going on.[38]

In subsequent presentations at numerous medical conferences, Dr. Kniker emphasized the importance of delayed-onset adverse reactions to food. And in a chapter in *Handbook of Food Allergies,* edited by James Breneman, M.D., he said that in the years to come it is likely that adverse reactions to foods will rival or surpass the importance of adverse reactions to inhalants in the fields of allergy and immunology.[39]

Other academicians, including Douglas Heiner, University of California (Torrance); Sami Bahna, University of South Florida; Walter W. Tunnessen, Jr., Children's Hospital of Philadelphia; and Frank Oski, Johns Hopkins Medical School, have expressed interest in this type of food sensitivities.

In his book published in the 1970s, *Don't Drink Your Milk,* Oski included a chapter, "Milk and the Tension-Fatigue Syndrome."[40] I was especially pleased to read this chapter as it was based almost entirely on my 1975 article, "Food Allergy—the Great Masquerader,"[41] published in Pediatric Clinics of North America.

Why You Need to Know About Food Allergies/Sensitivities

Many years ago, an observant mother convinced me (against my will!) that sensitivities to cow's milk caused her 12-year-old son to complain of headache, fatigue, abdominal pain and muscle aches.

A short time later, I came across articles by Drs. Albert Rowe[42] and Theron Randolph[43] published in the peer-reviewed medical literature discussing food allergies and sensitivities. They reported that many of their patients improved—often dramatically—when they avoided wheat, corn, milk, egg and other foods.

So I began putting a number of my tired, inattentive, irritable patients on a one-week elimination diet. Although I didn't help all of them, I was excited because I began to receive reports from mothers who said:

> Susie's like a different child. But when she drinks chocolate milk or eats corn chips, wheat or eggs, her symptoms return.

I collected and summarized my findings on 50 of these patients and published them in a major pediatric journal more than 30 years ago.[44] The diagnosis of food sensitivity was made in the following manner: Symptoms and signs were relieved by eliminating suspected foods from the diet for 5 to 12 days and then reproduced by giving the food back to the child, and noting reactions.

In gathering material for this book, I sought the help and consultation of many people. Included were colleagues who were knowledgeable and experienced in treating patients with yeast-related health problems and other colleagues who were not. One academic physician in the latter group asked, "Why are you including so much material on food allergies and sensitivities in your book?"

In responding, I cited the observations of a number of clinicians and researchers, including Dr. J. O. Hunter, a Fellow of the Royal College of Physicians, who, in a discussion of food intolerance in the British journal, *The Lancet,* suggested that patients with food intolerances have an abnormal gut flora even though pathogens may not be present. He said:

> If food tolerance is not an immunologic disease, but a disorder of bacterial fermentation in the colon, it might be more appropriately named an "enterometabolic disorder." This is of more than mere terminological importance: modern microbiology has opened the way to the manipulation of bacterial flora to allow the correction of food intolerances and thus the control of disease.[45]

Another European researcher and clinician, Gruia Ionescu, Ph.D.,[46] shared Hunter's views about the relationship of gut flora to food sensitivities. And in an interview with Marjorie Hurt Jones, R.N., editor of *Mastering Food Allergies,* Ionescu stated that he had found candida overgrowth in the gut in approximately half of his allergic patients.

Still other investigators, including Leo Galland, M.D.[47] and W. Allen Walker, professor of Pediatrics at Harvard Medical School, have also been studying the gut. Some of Walker's research studies have focused on the role of the mucosal barrier in handling allergens. In a comprehensive review of antigen handling by the gut, he commented:

> There's increasing experimental clinical evidence that suggests that large antigenically active molecules can penetrate the intestinal surface, not in sufficient quantities to be of nutritional importance, but in quantities that may be of immunologic importance.
>
> This observation could mean that the intestinal tract represents a potential site for the absorption of . . . ingested food antigens that normally exist in the intestinal lumen.[48]

Neither Hunter nor Walker mentions the possible role that candida overgrowth contributes to abnormal gut flora. Yet, the observations of Galland, Ionescu and other candida clinicians about the favorable response of individuals with food sensitivities to antiyeast medications and sugar-free special diets provide support for the yeast connection to food allergies and sensitivities.

Tracking Down Your Hidden Food Sensitivities

Almost without exception, every person with a yeast-related problem is bothered by food sensitivities. To identify the foods that may be contributing to your symptoms, you must carefully plan and properly execute an elimination/challenge diet. Here's an edited transcript of a tape-recorded visit with one of my patients that will tell you what to do and how to do it.

The diet is divided into two parts: First you'll eliminate a number of your usual foods to see if your symptoms improve or disappear. Then, after five to ten days, when your symptoms show convincing improvement, eat the eliminated foods again—one at a time—and see which ones cause the symptoms.

Keep a record of your symptoms in a notebook:

a. for three days (or more) before beginning the diet
b. while you're following the elimination part of the diet (five to ten days—occasionally longer)
c. while you're eating the eliminated foods again—one food per day.

During the first two to four days of the diet, you're apt to feel irritable, hungry and tired. You may develop a headache or leg cramps. If the foods you've avoided are causing your symptoms, you'll usually feel better by the fourth, fifth or sixth day of the diet. Almost always, you'll improve by the tenth day.

After you're certain that you feel better, and your improvement has lasted for at least two days, begin adding foods back to your diet—one at a time. If you're allergic to one or more of the eliminated foods, you'll usually develop headache, fatigue or other symptoms when you eat the foods again. These symptoms will usually appear within a few minutes to a few hours. However, sometimes you may not notice the symptoms until the next day.

On the diet you can eat any meat but bacon, sausage, hot dogs or luncheon meats; any vegetables but corn; any fruits but citrus. You can also eat rice, rice crackers, oatmeal and the grain alternatives, amaranth and quinoa (obtainable from health food stores), and nuts in shells or unprocessed nuts.

After you've completed the elimination part of your diet, add the following foods to your diet—one food per day. Make sure you add the foods in pure form. Here are suggestions.

- Egg: Hard-boiled, or scrambled in pure safflower or sunflower oil.
- Citrus: Peel an orange and eat it.
- Milk: Use whole milk.
- Wheat: Cream of Wheat or Shredded Wheat.

- Food coloring: Buy a set of McCormick's or French's dyes and colors. Put a half teaspoon of several colors in a glass. Add a teaspoon of that mixture to a glass of water and sip on it.
- Chocolate: Use Baker's cooking chocolate or Hershey's cocoa powder. You can sweeten it with liquid saccharin.
- Corn: Use fresh corn on the cob or pure corn syrup.
- Sugar: Use plain cane sugar.

Here are other suggestions:

- Plan ahead. Don't start your diet the week before Christmas or some other holiday.
- Don't start it when you're traveling or visiting friends or relatives.
- Add back first the foods you least suspect.
- Save until last the foods you suspect are troublemakers.
- Eat a small portion of the eliminated food for breakfast. If still no reaction, eat more of the food for lunch, for supper and between meals, too.
- Keep the rest of your diet the same while carrying out the challenges.
- If you think you develop symptoms when you add a food, but are uncertain, eat more of the food until your symptoms are obvious. But don't make yourself sick.
- If you show an obvious reaction after eating a food, don't eat more of that food. Wait until the reaction subsides (usually 24 to 48 hours) before you add another food.
- If a food really bothers you, you can shorten the reaction by taking two tablets of Alka Seltzer Gold dissolved in a glass of water. And to get the food out of your digestive tract, take a saline laxative or a big dose of milk of magnesia.
- If you find that a food causes your symptoms, keep it out of your diet for three to four weeks, then cautiously try it again. Many food-sensitive people find that they can eat a small amount of the food every four to seven days.
- If one or more foods cause problems and you're still having symptoms, try the "cave man diet." On this diet you'll eliminate beef, chicken, pork, white potato, tomato, rice, oats,

coffee, tea and any food or beverage you consume more than once a week. (You'll find detailed instructions for carrying out this diet in *The Yeast Connection and the Woman,* pages 512–520.)

References

1. Skolnick, A., "Even Air in the Home Is Not Entirely Free of Potential Pollutants: Medical News and Perspectives," *JAMA,* December 8, 1989; 262:3102–3103, 3107.

2. Hileman, B., "Multiple Chemical Sensitivity," *Chemical & Engineering News,* July 22, 1991; pp. 26–27. Copyright 1991, American Chemical Society.

3. Browder, S., "Are You Allergic to the Modern World?" *Woman's Day,* April 1, 1992.

4. Rea, William, Personal communication, 1983.

5. Lawson, L., *Staying Well in a Toxic World,* The Noble Press, Chicago, IL, 1993.

6. Ulene, A., and Ulene, V., The Vitamin Strategy, Ulysses Press, Berkeley, CA, 1994.

7. Carper, J., *Stop Aging Now!,* HarperCollins, New York, 1995.

8. Good, R., as cited in Crook, W. G., *The Yeast Connection,* Third Edition, Professional Books, Jackson, TN and Vintage Books, New York, 1986; p. 371.

9. Meydani, S., as quoted by Angier, N., "Vitamins Lend Support as Potent Agents of Health," *New York Times,* March 10, 1992.

10. Baldwin, R. S., *The Fungus Fighters,* Cornell University Press, Ithaca, New York, 1981.

11. Copyright, *Physicians' Desk Reference,* 46th Edition, Medical Economics, Montvale, NJ 07645, 1992; p. 605. (Reprinted by permission. All rights reserved.)

12. Como, J. A. and Dismukes, W. E., "Oral Azole Drugs As Systemic Antifungal Therapy," *N. Engl. J. Med.,* 1994; 330:263–272.

13. Grant, Susan M. and Clissold, S. P., "Itraconazole: A review of its pharmacodynamic and pharmacokinetic properties and therapeutic use in superficial and systemic mycoses," *Drugs,* 1989; 37:310–344.

14. Jacob, S. and Nall, L., "Discovering Antimycotic Drugs: Today and Tomorrow," *Cutis,* 1990; 45:245–250.

15. Murray, M. P., *Natural Alternatives to Over-the-Counter and Prescription Drugs,* Professional Physicians Publishing and Health Services, Inc., Houston, TX.

16. Murray, M. P., *The Healing Power of Herbs,* 2nd Edition, Prima Publishing, Rocklin, CA. 1995.

17. Farber, A. C., *Coll/Ag-30 The Silver Micro Bullet,* Professional Physicians Publishing and Health Services, Inc., Houston, TX, 1995.

18. Smith, C. W., Jr., "Exercise: Practical Treatment for the Patient with Depression and Chronic Fatigue," Primary Care, 1991; 18:2:271-281.

19. McIlhaney, J. S., Jr. with Nethery, S., *1250 Health-Care Questions Women Ask,* Baker Book House Company, Grand Rapids, MI, and Focus on the Family Publishing, Colorado Springs, CO, 1992; p. 816.

20. Gaby, A., *Preventing and Reversing Osteoporosis,* Prima Publishing, P.O. Box 1260 BK, Rocklin, CA 95677, pp. 219–224.

21. Burnstein, L., in comments to Kate Couric, The Today Show, NBC, September, 1994.

22. Ornish, D., *Dean Ornish's Book for Reversing Heart Disease,* Ballantine Books, 1990; pp. 85–103.

23. Jones, J., in the Boston Women's Health Collective, *The New Our Bodies, Ourselves,* Touchstone, Simon and Schuster, New York, 1998; pp. 65–78.

24. Lark, S. M., *PMS—Self-Help Book,* Celestial Arts, P.O. Box 7327, Berkeley, CA, 1984.

25. Haas, E. M., *Staying Healthy with Nutrition,* Celestial Arts, Berkeley, CA, 1992; pp. 487–497.

26. Cousins, N., "Anatomy of an Illness (as perceived by the patient)," *N. Engl. J. Med.,* 1976; 295:1458–1463.

27. Cousins, N., *Anatomy of an Illness,* Bantam Books, New York, 1981.

28. Cousins, N., *The Healing Heart,* Avon Books, New York, 1983.

29. Dossey, L., *Meaning and Medicine,* Bantam Books, New York, 1991.

30. Moyers, B., *Healing and the Mind,* Doubleday, New York, 1993.

31. Pelletier, K., in Goleman, D. and Gurin, J., *Mind/Body Medicine—How to Use Your Mind for Better Health,* Consumer Reports Books, Yonkers, NY, January 1993.

32. Northrup, C., *Women's Bodies, Women's Wisdom,* Bantam Books, New York, 1994; pp. 485–543.

33. Rochester, J., Montreal, Canada, Personal communication, 1994.

34. Von Pirquet, C., "Allergie," Munch Med. Wochenschr., 1906; 53:1457.

35. Speer, F., *Allergy of the Nervous System,* Charles C. Thomas, Springfield, IL, 1970.

36. Gerrard, J. W., Heiner, D. C., Ives, E. J. and Hardy, L. W., "Milk Allergy: Recognition, Natural History and Management," *Clinical Pediatrics,* 1963; 2:634.

37. Deamer, W. C., "Some impressions gained over a 37 year period," *Pediatrics,* 1971; 48:930.

38. Kniker, W. T., "Deciding the Future for the Practice of Allergy and Immunology," *Annals of Allergy,* 1985; 55:102.

39. Kniker, W. T., and Rodrigues, L. M., "Non-IgE-mediated and Delayed Adverse Reactions to Foods or Additives," in *Handbook of Food Allergies,* edited by Breneman, J. C., Marcel Decker, Inc., New York and Babel, pp. 125–147.

40. Oski, F., "Milk and the Tension Fatigue Syndrome" in *Don't Drink Your Milk,* Teach Services, Donivan Rd., Rt. 1, Box 182, Brushton, NY 12916.

41. Rowe, A. H., Sr., "Allergic Toxemia and Migraine Due to Food Allergy," *California and Western Medicine,* 1930; 33:785.

42. Crook, W. G., "Food Allergy—The Great Masquerader," *Ped. Clin of N. Amer.* 1975; 22:219–226.

43. Randolph, T. G., "Allergy as a Causative Factor of the Fatigue, Irritability and Behavior Problems of Children," *J. of Ped.,* 1947; 31:560.

44. Crook, W. G., et al, "Systemic Manifestations Due to Allergy," 1961; *Pediatrics,* 27:790–799.

45. Hunter, J. O., "Food allergy—or enterometabolic disorder?" *The Lancet,* 1991; 338:495–496.

46. Ionescu, G., As quoted by Jones, M., *Mastering Food Allergies,* Coeur d'Alene, ID, 1991.

47. Galland, L., "The Effect of Microbes on Systemic Immunity," in Jenkins, R. and Mowbray, *Post-viral Fatigue Syndrome,* John Wylie and Sons, 1991.

48. Walker, W. A., in Brostoff and Challacombe, *Food Allergy and Intolerance,* London, Balliere Tindall, Philadelphia, W. B. Saunders, 1987; pp. 209–222.

CHAPTER · TEN

More Information About Diets and Antiyeast Medications

■ ■ ■

Answers to Common Questions

Q Please tell me about the diet I will need to follow. As I look at your instructions I feel overwhelmed—even intimidated.

A I can understand your feelings. Eating, like breathing, is something we do naturally a number of times a day. It usually makes us feel good. In answering your question, I'd like to pass along the advice of the experienced candida clinician, George Kroker, M.D., who tells his patients:

> The diet itself is a "process"; it is not a project that can be "accomplished" in one visit. We will work on it, build on it, modify it and change it with the ultimate goal of improving your health.

I'd also like to point out that *restricted diets do not last forever.* As your health improves and you get the "yeast beast" under control, you can nearly always eat some of the foods that initially you need to avoid.

Q How do I get started?

A Read and re-read the instructions in the diet section of Chapter 9. Then go to your local health food store or food co-op. Personnel there will help you find the items you need. You'll also need a good cookbook to use as a reference source.

On your beginning diet, you'll need to avoid sugar and foods that are loaded with yeast. Yet, you can eat vegetables, lean meats, chicken, fish, whole grains and the grain alternatives.

Q **Please tell me more about food yeasts.**

A Yeasts and molds are found on the surfaces of all fruits, vegetables and grains. The yeast we deal with in most foods and beverages are the saccharomyces. They're found in beer, wine and breads, as well as dried, primary-grown, baker's yeast. These nutritional yeasts are an excellent source of enzymes, B vitamins and amino acids. Many people with candida-related health problems may not need to avoid them. Moreover, completely yeast-free diets are impossible to come by.

Q **How will I know if yeast-containing foods bother me?**

A Do a yeast "challenge." Here's how it works. After you've avoided all yeasty foods for ten days, eat one of them when you aren't changing anything else in your treatment program. This will usually give you the answer. If you're sensitive to yeast, you'll develop symptoms within a few minutes to 24 hours. In discussing this subject, Sidney Baker, M.D., said,

> Remember that once you have already decided that you are sensitive to yeast, you will need to be your own judge as to how much you tolerate various foods that may contain some yeasts or molds.

Q **Suppose I flunk the yeast challenge? What then? Will I remain sensitive?**

A I don't know. Many people lose their sensitivity to yeast and other foods as the candida yeasts in their intestinal tract decrease in number. So you might repeat your yeast challenge every one to two months. If you still develop symptoms, you may be desensitized to food yeasts by a physician who is a member of the American Academy of Environmental Medicine. (See page 237.)

Q What about fruits?

A After you've finished the yeast challenge, and have been on the diet for about three weeks, you can experiment with fruits if you're doing well. But do it cautiously with only a small portion once a day. Then you can feel your way along.

How often and how much fruit you can take is an individual matter. After a month or so, some people seem to be able to eat one or two fruits every day without experiencing problems. Yet, in an occasional person, fruits must be avoided completely for many months. To determine your own tolerance—just as you do with yeast—you *experiment.*

One more word of caution about fruits, especially if you're yeast sensitive. Avoid the juices—except for freshly squeezed juice—because they're loaded with yeast.

Q I now have a better understanding about yeasts and fruits. Now I'd like to talk about sugar because I crave it.

A Go *very* slow in trying the simple sugars. Wait many weeks before you experiment. I'm talking not only about the "white stuff," but also about corn syrup, honey and maple syrup. They all feed the yeast.

Q Will I ever be able to eat sugar again?

A Yes, but in limited quantities after you've really gotten your health and your life back on track. As you've read elsewhere in this book, sugar really promotes the multiplication of yeasts in the digestive tract. You might compare it to putting kerosene on the glowing embers of a barbecue grill.

Sugar isn't a nutritious food. It contains only calories with no vitamins, minerals, fiber or other health promoting nutrients.

Some 200 years ago, our ancestors consumed about seven pounds of sugar a year. They used it as a condiment.

By contrast, Americans consume some 120 to 140 pounds a year—about 5 or 6 ounces a day. So wait until you're really feeling well before you experiment. To really enjoy good health, your sugar intake should resemble that of President George Washington and Benjamin Franklin.

Q What can I use in the place of sugar in my coffee, tea or foods?

A To sweeten your beverages, try the South American herb, stevia. You can find it in most health food stores. For use on cereals and cookie recipes, try Jerusalem artichoke flour.

Q How about the artificial sweeteners?

A The main ones available today are saccharin and NutraSweet (aspartame). Based on research studies I've read recently, saccharin is safe. Moreover, I use it and recommend it without hesitation—in limited amounts. If you use it, get the liquid preparations because they do not contain added sugar.

I do not recommend NutraSweet because it causes adverse reactions in many people. (See page 183.) However, if you wish to test it at home, you can add a packet or two to a plain glass of water and see if you can tolerate it. Quoting Dr. Kroker again,

> The question of NutraSweet keeps coming up in my office. Although I don't recommend it for home use, it may be the "lesser of two evils" for people who are "on the road."

Q Will I need to take antifungal medications forever?

A No. As your health improves, your own immune system will control candida overgrowth in your intestinal tract or vagina. And there are many things you can do to strengthen your immune system. *Eating a more nutritious diet is an important first step.* Eat at least five vegetables a day because they are really "good for you." Research studies coming in from many different sources during the past decade show that broccoli, cabbage, carrots, yams, tomatoes and many others are loaded with *phytochemicals.*

Fruits also contain these important nutrients and—as we've already discussed—you can probably eat more fruits as the months go by. Many other measures I've discussed in this book also will help strengthen your immune system. These include limiting your exposure to toxic pollutants, taking supplemental vitamins and minerals, flax seed oil and probiotics. I also recommend other antioxidants, including CoQ_{10} and the bioflavonols.

Some people with mild to moderate yeast-related problems improve significantly—even dramatically—on nystatin and diet. And after one to three months, they can start tapering off the nystatin and take only acidophilus, garlic or other nonprescription antiyeast substances.

But other people with moderate to severe yeast-related problems may require Diflucan, Sporanox, Nizoral or Lamisil for several weeks. Then, as they improve, their physicians shift them over to nystatin or one of the nonprescription antifungals, including caprylic acid or grapefruit seed extract. *Each person is different and no two physicians follow the same treatment program.* Several of the physicians I've interviewed tell me they give their patients 100 milligrams of Diflucan, once or twice a week for many months. And in women, they are apt to prescribe extra antifungals the week before menstruation.

Q **Suppose I'm one of those people who improve dramatically on the yeast-free, sugar-free diet and antifungal medication. When I deviate from my diet or when I stop my antifungal medication will my symptoms come back?**

That's a good question. I'd again like to cite the comments of Dr. Kroker, who said:

> Candida-Related Complex often is associated with other disorders, most notably mold allergy, chemical sensitivities, autoimmune thyroiditis and food intolerances. Unless these illnesses are screened for, the treatment for candida (antifungal medicine and

> diet) may completely fail to ameliorate the patient's symptoms. I cannot overemphasize the importance of the "total load" in dealing with these patients.[1]

If your health problems are severe and of long duration, your physician may recommend that you take a prescription antifungal medication for 6 to 12 months, or even for several years.

Q In spite of what you just said, do you have words of encouragement?

A Yes. I base my answer on the reports I've received from hundreds of my own patients. Here is an example:

In *The Yeast Connection* I told the story of 33-year-old Janet who was bothered by symptoms which affected just about any and every part of her body. These included headaches, fatigue, dizziness, puffiness, excessive weight gain, sexual dysfunction, bladder problems, chemical sensitivities and much more. She was truly "sick all over."

Three months after starting on a comprehensive treatment program, she reported that she was symptom free and well except when she cheated on her diet. Although Janet moved to another state in the mid-80s, I've kept in touch with her through occasional letters and phone calls.

To get an update, I wrote her in March 1993 and asked her to send me a progress report. Here are excerpts from her letter:

> I'm doing quite well, enjoying my forties healthwise much more than my thirties. I'm 43 now and I'm quite busy with three children. One almost 17, one 14 and one 5. I operate with my energy level much higher today than when we first began the "yeast journey" together.
>
> I still don't eat bread, but I do have yeast in my diet, some sugar and other junk. But overall my eating is much healthier than 10 years ago.

Q I know that antibiotics such as amoxicillin, ampicillin and Keflex make yeasts multiply. But, should I stay away from these drugs?

A Yes and no.

First to explain the "yes" answer. Most minor respiratory infections are caused by viruses and presently available antibiotics are ineffective against them. So if you have a mild fever, cough and cold, earache or sinus congestion, don't rush to your physician asking for an antibiotic. Instead, wait it out. Take big doses of vitamin C, zinc sulfate 15 milligrams, three or four times a day and the herb echinacea. Drink lots of fluids and consume a nutritious low sugar diet. Let your own immune system help you get well.

If you're really sick, however, by all means see your physician. Let him examine you and determine whether you need an antibiotic for a bacterial infection. If he gives you an antibiotic, also ask him to prescribe nystatin. I recommend one 500,000 unit or one-eighth teaspoon of nystatin powder three times daily. In addition, take a probiotic which contains *Lactobacillus acidophilus* and *Bifidobacterium bifidum.* In so doing, you keep good bacteria in your intestinal tract. In my practice, I recommend that the nystatin be continued for a week—and sometimes for many weeks or months—and the same goes for the probiotic.

Q My husband's health is generally good, certainly compared to mine. He works every day and has few complaints. Yet, he's taken antibiotics on several occasions during the past five years because of sinusitis. Should he be treated?

A From what you tell me, I feel that some of his symptoms may be yeast connected, and that a trial of the diet and antifungal medication could make him feel better. And even in men who are symptom-free, a yeast-control program will often help the spouse or sexual partner.

Several years ago, 40 women with recurring vaginal yeast infections were recruited for participation in a double-blind study. Each woman received 100 milligrams of Sporanox daily for five days. The sexual partners of one-half of the women were given Sporanox, while the other half were given a placebo.

After a month, 14 percent of the women whose sexual partners received the placebo, had another vaginal yeast infection; *none of those whose sexual partners had been treated had a recurrence.* Based on this report and observations of many candida clinicians, antiyeast treatments should be a family affair.

Q My family physician is a nice person. I've gone to him for many years. I like him. But he doesn't know about yeast-related problems, he doesn't understand and he is skeptical. What do I tell him?

A No one answer will suit every situation. Nevertheless, here are suggestions. First, thank your physician for his kindness and interest and for the care he has given you. But then do not hesitate to tell him about your experiences—especially about your improvement on dietary changes and nonprescription antifungal medications and/or nutritional supplements. Depending on the response you receive, you can move on from there and provide him with additional information.

If he or she is skeptical, but willing to listen, write to the International Health Foundation, Box 3494, Jackson, TN 38303, for additional information. (See pages 238–240.)

How I Manage My Yeast Patients

During the past five years, I've "picked the brains" of several dozen physicians who are knowledgeable and experienced in treating patients with yeast-related disorders. I've also received countless letters and phone calls from people with these disorders. Although I've acquired a lot of information, every day I learn something I did not know before.

Recently a physician wrote me and said, "Until recently I had been skeptical of 'the yeast connection.' But after one of my chronically complaining patients responded dramatically to diet and antifungal medication, I'd like to know more. Please tell me how *you* manage *your* patients."

Here's what I told him:

The health problems, family and work situations of each patient are unique and differ from patient to patient. Yet, here's the approach I use in most adults I see in consultation . . . and most, but not all of them are women.

Before the first visit, I send prospective patients the long yeast questionnaire and information about my office, including charges that can be expected. I ask them to write me a long letter telling me their medical, personal and work histories right from the beginning. As you might guess, these letters are sometimes 10 to 15 pages long. I also review pertinent medical records sent by the referring physician.

The first visit is designed to "get acquainted." I encourage patients to come with another family member or friend to provide support as they begin the journey to regain their health. I review and expand on the medical history, personal history, social history, family history, usual diet and the environmental exposures at home or at work. I also ask about nutritional supplements, and much more.

Although I recommend getting rid of junk food and cutting down on sugar, I never put patients on a strict diet at the time of our first visit. Yet, we do talk about diets and I ask them to review the diet section of this book and begin to make plans. And, as you might guess, I also recommend my comprehensive 1995 book, *The Yeast Connection and the Woman*—even though the patient may be a man, a young child or a teenager.

During the visit I point out to my patients that cigarette smoke and indoor odorous chemicals play a part in making many people sick. And I ask them to start banishing these pollutants. I also recommend a comprehensive multivitamin mineral program—flax seed oil and probiotics, including *Lactobacillus acidophilus* and *Bifidobacterium bifidum.* A member of my office staff and I also answer questions.

I conclude the visit by saying, "We've made a beginning today and I feel you can be helped. At your next visit, in a week or ten days, I'll answer your questions and we'll move ahead with the diet.

I rarely, if ever, start my patients on antifungal medications until they have been on a sugar-free diet for at least a week. Here's why: The intestinal tract of an individual, whose diet is loaded with sugar and other simple carbohydrates, will be loaded with yeasts. And when they take an antifungal medication, they will usually experience severe "die off" reactions. (See page 122.)

After the patient follows the strict diet for a week, I usually prescribe 200 milligrams of Diflucan once a day for three days, then 100 milligrams a day. If it is well tolerated (as it usually is), after two weeks I add nystatin, 1,000,000 units, four times a day.

If after three weeks of antifungal therapy, the patient continues to improve, I may reduce the dose of Diflucan to 100 milligrams every other day for a week and then discontinue it while continuing with the nystatin.

I'd like to again emphasize that each patient is different and these notes provide only a general outline of my treatment program for most adults. In patients with more serious problems, such as multiple sclerosis or severe depression, I may prescribe the Diflucan sooner and give 200 milligrams a day for one to four months.

If the Diflucan and/or nystatin do not seem to be helping, I often change to Sporanox or Nizoral. And in some patients, I recommend nonprescription antiyeast medications, including garlic, caprylic acid, citrus seed extract or Tanalbit (see pages 128–133).

In all of my patients, including those who are doing well, I encourage them to become life-long students. I've listed a number of newsletters, magazines and books that I recommend in the Appendix.

I also tell them that people who eat less meat and more plant foods will enjoy better health than their contemporaries. I also recommend CoQ_{10}, grape seed extract, *Ginkgo biloba* and other nutritional supplements.

In conclusion, I always emphasize that while diet and antiyeast medications help, they do not provide a quick fix for yeast-related health problems. Patients who fail to improve may resemble the "overloaded camel" and other bundles of straw may be loading them down.

These may include tobacco smoke, toxic carpets and/or insecticides in their homes and work places. Intestinal parasites or hidden food sensitivities may also be contributing to their problems. So may unusual stresses at home or elsewhere.

When one of my yeast patients continues to experience problems, I schedule an in-depth consultation visit. I review the history and every part of the treatment program. In many patients, I order a comprehensive examination of the stool to detect parasites and/or other abnormalities. And in some patients I order blood tests to help pinpoint food sensitivities and identify nutritional deficiencies.

References

1. Kroker, G. F., in Crook, W. G., *The Yeast Connection and the Woman,* Professional Books, 1995: p. 654.

2. Calderon-Marques, "Itraconazole in the Treatment of Vaginal Candidosis and the Effect of Treatment of the Sexual Partner," *Review of Infectious Diseases,* Vol. 9, Supplement 1, Jan/Feb, 1987. University of Chicago, pp. 143–145.

CHAPTER · ELEVEN

Other Therapies That May Help

■ ■ ■

Thyroid Hormones

Thyroid Hormone Deficiency (Hypothyroidism)

Today if a thyroid deficiency is suspected, blood tests to measure T3, T4 and TSH are usually ordered. When these tests are normal, most physicians conclude that thyroid supplements aren't needed. Yet, other physicians feel that thyroid deficiency occurs commonly in people with fatigue and other generalized symptoms—even if their blood tests are normal.

Fifty years ago, Broda O. Barnes, M.D., began writing and talking about thyroid deficiencies. In an article published in *The Journal of the American Medical Association*, he described the use of basal temperatures in determining such deficiencies.[1]

Over the next several decades, Dr. Barnes published a number of reports on the value of thyroid supplements in treating health conditions ranging from menstrual disorders to heart disease.

In 1976 he published a book entitled *Hypothyroidism: The Unsuspected Illness.*[2] In this book, Dr. Barnes described his success in helping many of his patients with thyroid supplements even though their laboratory studies showed no evidence of hypothyroidism.*

---

*You can obtain a copy of Dr. Barnes' book and other information by writing to the Broda O. Barnes, M.D., Research Foundation Inc., P.O. Box 98, Trumbull, CT 06611–0098; phone number: 203-261-2101, Fax: 203-261-3017.

In the early 1980s, a medical colleague, who shared my interest in food and chemical sensitivities and yeast-related health problems, said:

> I found that some of my yeast patients are low in thyroid. You ought to try Dr. Broda O. Barnes' basal temperature test on some of your difficult patients. As you may know, he found that when a person's underarm temperature is consistently under 97.8 degrees before they get out of the bed in the morning, they are usually deficient in thyroid. Supplemental thyroid hormone will provide them with a lot of help.

I began to look for thyroid deficiency and was delighted at the response of many of my patients to thyroid supplements. I made a brief reference to Dr. Barnes' work in the third edition of *The Yeast Connection,*[3] published in 1986, and I included a comprehensive discussion in *The Yeast Connection and the Woman* (4th printing, 1998).[4]

Observations of Experienced Clinicians

During the 1990's, I interviewed a number of physicians including Drs. Alan Gaby, Keith Sehnert, Jorge Flechas and Ken Gerdes. These experienced clinicians told me that thyroid supplements provided significant help for many of their patients with yeast-related disorders. Sehnert pointed out that many of his patients with candida problems are also troubled by food allergies and hypothyroidism. He noted that individuals with cold hands, cold feet and fatigue are especially apt to be deficient in thyroid. And Gerdes, a Denver physician commented:

> Following the lead of the late Dr. Broda Barnes, I use the basal temperature. Yet, rather than having the patients take their temperature under the arm, I ask them to measure it under their tongue for five minutes . . . If a patient's temperature is consistently under 97.4 or 97.6 in the morning, I often prescribe T3 (in addition to Armour thyroid tablets) and it seems to play an important role in helping my patients get better.

Adrenal Hormones

Since the early 1950s, cortisone, prednisone and related steroid drugs have helped relieve the symptoms in people with asthma, hives and other disorders. Yet, when these drugs were given in customary large doses, they caused serious adverse effects. So most physicians felt that they should not be given over many weeks or months except in exceptional situations.

Cortisol

For over 30 years, William McK. Jefferies, a physician with impeccable academic credentials, has studied mild adrenocortical deficiency and the use of small doses of cortisol. Moreover, he found that such doses are safe and effective in patients with a variety of common complaints, including rheumatoid arthritis, allergic rhinitis, chronic fatigue syndrome, asthma and diabetes. In 1981, he published a book, *Safe Uses of Cortisone*[5] and in a 1994 paper, he said:

> Cortisol (hydrocortisone) is a normal hormone which is essential for life in humans. . . . In the treatment of adrenal deficiency in our practice . . . dosages of 2.5 milligrams to 7.5 milligrams, four times daily, are satisfactory maintenance dosages, depending upon the degree of the deficiency. . . . Patients have been treated with this schedule of cortisol or cortisone acetate for as long as 40 years without significant problems. . . . There is . . . no reason to fear that physiologic [small] doses of cortisol will produce any of the harmful side effects of pharmacological doses.[6]

DHEA (Dehydroepiandrosterone)*

In the February 1994 issue of *Health & Healing*, Dr. Julian Whitaker said:

> DHEA is the "mother" hormone produced by the adrenal gland. Your body readily converts it on demand into active hormones such as estrogen (and progesterone) . . . In addition, DHEA decline signals age-related disease.

In his continuing discussion, Dr. Whitaker cited research studies by Elizabeth Barrett-Conner, M.D., of the University of Califor-

---

*For a list of DHEA references, send a long, self-addressed, stamped envelope to Phillips Publishing, Customer Service-DHEA, 7811 Montrose Rd., Potomac, MD 20854. See also pages 257–258.

nia School of Medicine in San Diego, who studied 5,000 women. She found that those who developed breast cancer had subnormal urinary excretion of DHEA metabolites as long as nine years before the development of the disease.

He also cited the research studies of Eugene Roberts, M.D., who found that elderly volunteers with moderate memory loss, who took DHEA supplements, scored higher on two of the four measurements than those who received a placebo. And he commented:

> DHEA blood levels are easily measured, and I often prescribe supplementation to bring a patient's blood levels up to a healthier level of 20- to 30-year olds. I am surprised at how low the blood levels of DHEA are in some patients who are ill with heart disease, diabetes or cancer. In addition, it also surprises me that so few doctors measure DHEA blood levels and prescribe supplementation.
>
> There is no patent on DHEA, so no drug company is interested in promoting it as therapy. Consequently, it languishes on the shelf and many doctors are under the impression it is illegal. Though it is not "approved" by the FDA for any specific medical condition, it is without question legal. Any doctor can prescribe it . . . I've been using DHEA in my patients for some time now, with sometimes startling results.

My Comment

Many people with yeast-related disorders and other chronic health problems show grossly abnormal DHEA levels. These levels can be determined by a DHEA blood test, which can be ordered by your physician. Although DHEA can be obtained without a prescription, I obtain the blood test before recommending DHEA to my patients.

CoQ_{10}

CoQ_{10}, also known as ubiquinone, is a nutrient that was first extracted from beef heart mitochondria by scientist F. L. Crane and his group in the United States in 1957. A great deal of the work on this substance has been carried out since that time by Dr. Karl Folkers and research colleagues at the University of Texas in Austin. It also has been researched and used extensively in

Japan, where 252 commercial preparations of CoQ_{10} are supplied by more than 80 pharmaceutical companies. According to the authors of *The Miracle Nutrient, Coenzyme Q_{10}*:

> On April 14, 1986, Karl Folkers was honored with the Priestley Medal, the highest award bestowed by the American Chemical Society in recognition of superior accomplishments in chemistry and medicine. It was presented to Dr. Folkers in recognition of his work with Coenzyme Q_{10}, vitamin B_6 and B_{12}.

The book reviews many reports that describe the value of Coenzyme Q_{10} in people with congestive heart failure and other types of heart disease. It cited other studies, which show that CoQ_{10} boosts the performance of immune system cells. The authors of the book also concluded that research into CoQ_{10} and the immune system has received less emphasis and publicity than its role in treating heart disease. They said:

> Our research with animal models, and the studies of other scientists, have proved conclusively that CoQ_{10} can produce a profoundly beneficial effect on the immune system . . . CoQ_{10} displays no toxic effects whatsoever . . . Clinical studies, under the auspices of the FDA, show that CoQ_{10} is much safer than many drugs presently on the market.[7]

Support for CoQ_{10}

I take CoQ_{10} every day as a nutritional supplement. So do other members of my family. During the past seven years, I also have used CoQ_{10} as one part of a comprehensive program in treating my patients with yeast-related disorders. Here's the story on one of my patients.

> Bobbie, a 46-year-old woman with recurrent asthma, sinusitis, rhinitis, fatigue, headache and other symptoms, had failed to improve on repeated courses of antibiotic drugs and bronchial dilators. On a comprehensive treatment program, which included an improved diet, vitamin and mineral supplements, small doses of nystatin and CoQ_{10}, she showed remarkable improvement in her health status. In a recent letter to me she told me that she was enjoying better health than she could remember at any time

in her life. She said, "I believe CoQ_{10} has certainly contributed to my ability to stay well."

A Report from the Peer-Reviewed Medical Literature

In an article in the *Journal of Clinical Pharmacology,* W. H. Frishman, Department of Medicine at Mt. Sinai Hospital and Medical School, said this about CoQ_{10}:

> The substance has been used in oral form to treat various cardiovascular disorders, including angina pectoris, hypertension and congestive heart failure. Its clinical importance is now being established in clinical trials worldwide.[8]

Other Reports on the Value of CoQ_{10}

In his newsletter, *Health and Healing,* Dr. Julian Whitaker described the response of his patient, Charmaine:

> When I first met her four years ago, she was as near death as any patient I had ever seen. On a comprehensive nutritional regimen, which included large doses of CoQ_{10}, her response was almost immediate and miraculous.
>
> Today, Charmaine takes no heart medications at all and leads an active life. She does, however, continue to take CoQ_{10} as well as other vitamins and minerals.
>
> CoQ_{10} is also a potent antioxidant protecting cells from free radicals, the by-products of energy production . . . When CoQ_{10} is low, nothing in your body works well.[9]

I read another exciting report about the value of CoQ_{10} in a research review by Alan R. Gaby, M.D.[10] He cited a recent article describing the response of four women with metastatic breast cancer who were treated with CoQ_{10} in large doses. One of the patients was a 44-year-old woman whose breast cancer metastasized (spread) to numerous places in the liver. After taking CoQ_{10} for eleven months, the patient was reported to be "in excellent health."

Gaby reported a story on another woman with breast cancer, which had spread to her liver and lungs. After six months of CoQ_{10} therapy, this patient, too, was reported to be in excellent health.[11]

From other sources I learned that the dose of CoQ_{10} in these patients with cancer was approximately 400 milligrams a day. Based on this report (and others), if I developed cancer, heart disease or an immune disorder of any type, I'd load up on CoQ_{10}, a nutritious vegetarian diet and take many nutritional supplements.

Vitamin C

If you're like most North Americans, you've read and heard about the double Nobel Prize winner, Linus Pauling, Ph.D. Twenty-five years ago he wrote a book entitled *Vitamin C and the Common Cold.* Then in 1986, he published another book, *How to Live Longer and Feel Better.* In the latter book, Pauling told the fascinating story of Dr. James Lind, a physician in the British Navy. In the mid 1700s Lind found that limes (and other fresh fruits) kept sailors from developing scurvy.

Yet, it wasn't until the 1930s that the Hungarian researcher, Dr. Albert Szent-Györgyi, learned why lime juice worked. It contained vitamin C and Györgyi received the Nobel Prize for his discoveries.

In his books and lectures, Pauling recommended doses of vitamin C that were 10, 100 or 1,000 times greater than the amount needed to prevent scurvy. Moreover, many scientific reports have shown that vitamin C is an important antioxidant, which plays a major role in strengthening the immune system.

How much vitamin C should you take and how do you determine the proper dose? The observations of Dr. Robert Cathcart help answer these questions. This internationally known orthopedic surgeon was troubled by frequent colds and constant hay fever. After reading Pauling's first book, Cathcart began taking one teaspoon of ascorbic acid powder in water four times a day. *He was astonished to find that his hay fever symptoms were abolished as long as he kept taking the high doses of vitamin C.*

Subsequently, Cathcart found that large doses of vitamin C helped his patients with fractures and other injuries heal more rapidly and with much less pain. So he began to look for further information and he found a number of reports in the medical literature documenting the value and safety of vitamin C therapy.[12,13,14,15]

Cathcart soon found that other patients with various health problems, who took *huge* doses of Vitamin C, improved more rapidly. Some 20 years ago, in a letter to the editor in the *Medical Tribune* (June 25, 1975), he reported that the "bowel tolerance" to ascorbic acid (vitamin C) of a person with a healthy gastrointestinal tract was somewhat proportional to the toxicity of their disease.

In two articles published in the medical literature, Cathcart defined the bowel-tolerance dose as the amount of ascorbic acid tolerated orally that almost, but not quite, causes a marked loosening of the stools.[16, 17]

In his further work with his patients, Cathcart found that healthy people could tolerate orally 10 to 15 grams of ascorbic acid in a 24-hour period. Yet, sick people, including those with influenza or mononucleosis, could take 200 or more grams in 24 hours.

In determining the appropriate dose, he advises his patients to begin with hourly doses of ascorbic acid powder dissolved in small amounts of water. Later, after the patient has learned to accurately estimate the proper dose, comparable doses of ascorbic acid tablets or capsules are used.

My Comments

During the past decade many of my own patients, as well as people who have written and called me, have confirmed Dr. Cathcart's observations. I've found it especially helpful in people with multiple chemical sensitivities. Vitamin C is only one of the many antioxidants and other nutritional supplements that I recommend. Here are instructions I've given to many of my patients with chronic health problems:

- Go to your health food store and purchase several ounces of vitamin C powder or crystals. Add a teaspoon (4 grams) to a pint of water. Drink a couple of ounces every hour or two throughout the day, unless it upsets your digestive system.

- Continue this practice for several days and, if it doesn't bother you in any way, add a second teaspoon of vitamin C powder and continue the dose.
- If you have a cold or you've been exposed to chemical pollutants, try even larger doses of vitamin C, as recommended by Dr. Cathcart.

Vitamin B_{12}

If you're bothered by fatigue, high doses of vitamin B_{12} may help you. I base that statement on my own experiences, plus reports by others that this safe, inexpensive vitamin is often highly effective by mouth as well as by injection.

For more than 20 years, H. L. Newbold, M.D., a New York physician, has been writing and talking about the beneficial effects of vitamin B_{12} injections in the treatment of fatigue, backache, depression, poor memory and other health problems.[18] Recent reports in the *New England Journal of Medicine* stated that more than one fourth of a group of patients with neuropsychiatric abnormalities responded to therapeutic doses of B_{12}, even though there was no evidence of pernicious anemia. The symptoms in these patients included numbness, tingling, sensory loss, and dementia and other psychiatric symptoms.[19]

Medicines from Plants

Although plant-derived medicines were rarely mentioned in medical journals or conferences in the 1980s, during the mid-1990s they have received increasing attention. I found evidence of such attention in a two and one-half-page editorial published in the June 18, 1994, issue of the distinguished medical journal, *The Lancet*. Here are excerpts from "Pharmaceuticals from plants: great potential, few funds":

> Plants have long been a source of medicine . . . use of the herb quing hao, now a recognized source of antimalarial agents, was recorded for haemorrhoids, in 168 B.C. in China. In the UK and in North America, almost 25 percent of the active compo-

nents of currently prescribed medicines were first identified in higher plants.

The editorial told of many herbal remedies, including ginseng, echinacea, garlic and *Ginkgo biloba.* And in discussing the latter plant medication, the editorial said:

> Studies of some of these plants have yielded compounds with unique activity—e.g., ginkgolides, specific platelet activating factor antagonists that were obtained from the Chinese tree *Ginkgo biloba.* A standardized extract of ginkgo leaves is one of the most frequently prescribed medicines in Germany and is taken to alleviate cerebral ischaemia [inadequate blood flow].[20]

Bioflavanols

During the 1980s, I occasionally read or heard something about these nutritional substances, but I didn't really know much about them. Then, in June 1993, Bert Schwitters, a journalist I met in Holland, sent me an autographed copy of his brand new book, *OPC in Practice—Bioflavanols and Their Application.*[21] This book was written in collaboration with Professor Jack Masquelier, a distinguished French biochemist and a member of the medical faculty of Bordeaux University.

Professor Masquelier first isolated OPC from plant sources in 1948. In 1950, it entered the field of therapeutics as a vascular protector. About 30 years later (1979), Masquelier coined the term "pycnogenols" to make a clear differentiation between the non-bioavailable bioflavonoids and his bioflavanols.

What are the benefits of these plant products? According to a nutritional authority, Richard A. Passwater, Ph.D., these plant products scavenge free radicals and, in so doing, they can help you live better and longer and stay healthier. Moreover, they help protect you from about 80 diseases, including heart disease, cancer, arthritis and most other non-germ diseases that are linked to the deleterious chemical action of free radicals.

In his discussion of these plant-derived substances, Passwater said that they are remarkably safe nutrients that correct conditions because they strengthen capillaries, nourish the skin and balance histamine production.[22]

Grape Seed Extract

These plant products can be found in health food stores in two main forms sometimes called OPCs and at other times they're called PCOs (procyanidolic oligomers). In discussing them in his book, *The Healing Power of Herbs,* Dr. Michael Murray said:

> PCOs from both grape seeds and pine bark have been marketed in France for decades. (Sales for the grape seed extract in France are roughly 400 times greater than those for pine bark . . .) Although both sources can be used interchangeably . . . PCOs extracted from grape seeds have emerged as the preferred source. . . . It is far more economical to extract PCO from grape seed than it is from pine bark. As a result, the grape seed extract provides greater value at a lower price. . . .
>
> Regardless of the source, PCO extracts can be used to support good health. As a preventive measure and as antioxidant support, a daily dose of 50 milligrams of either the grape seed or pine bark extract is suitable. When used for therapeutic purposes, the daily dose should be increased to 150 to 300 milligrams. . . . PCO extracts exert no side effects.[23]

To learn more about these important plant products, get a copy of the Murray, Passwater and Schwitters books at your local health food store.

Magnesium

Many people, especially women, take calcium supplements in the hope of preventing osteoporosis and other disorders. Calcium is important—but magnesium is of equal or greater importance.

According to Dr. Sidney M. Baker, "Magnesium deficiency is widespread. The average daily need for magnesium in adults is between 500 and 1000 milligrams and a lot of people simply aren't taking that much. For my patients I recommend oral magnesium chloride."[24]

A former colleague of Dr. Baker, New York internist, Leo Galland, M.D., in discussing magnesium, said:

> The richest sources of magnesium are also the richest dietary sources of essential fatty acids . . . seed foods (including the whole grains, nuts and beans) and seafoods. Other foods, which are

relatively rich in magnesium, include buckwheat, baking chocolate, cotton seed, tea, whole wheat, collard greens, parsley and other leafy green vegetables. Magnesium is also plentiful in seafood, meats and fruit. What's more, you can protect your magnesium storers by avoiding the "magnesium wasters"; saturated fats and soft drinks, especially those containing caffeine.[25]

Enzymes

Michael McCann, M.D., Parma, Ohio, described his use of a pancreatic enzyme supplement in patients with multiple food allergies at a Food Allergy Symposium, sponsored by the American College of Allergy and Immunology.

In an abstract of his presentation, Dr. McCann described his experiences in treating a 37-year-old woman with lifelong eczema, intermittent diarrhea and weight loss.

This patient was given two to four capsules of a pancreatic enzyme supplement (25,000u protease/per capsule) before each meal. *She showed a complete clinical remission and resolution of eczema for the first time in her adult life. She also was able to discontinue all other drugs.*

Dr. McCann concluded that breaking up the allergens in food proteins by the pancreatic supplement lessened the absorption of large allergenic food particles (polypeptides) and, thereby, decreased the symptoms in food allergic patients.

In a 1999 conversation, Dr. McCann told me that Pancrease only helps food sensitive patients who show significantly positive skin tests. And, he said, it does not seem to work in patients with negative skin tests and other kinds of food intolerance.

R. W. Noble, M.D., of Dallas, Texas, also has found that pancreatic enzymes help his patients. Here are excerpts from a letter I received from him:

> I've used the enzyme Pancrease frequently, and at times Pan-Five, for people with food allergies, and have been pleased with their response. Many of these patients were troubled with digestive upsets, such as excessive gas. It also relieved these symptoms as well.
>
> Although I've used these enzymes primarily for digestive complaints, several of my patients have reported that they felt

better and showed improvement in other food-induced symptoms while taking these enzymes.

Hugging

In a chapter of their best-selling book, *Chicken Soup for the Soul—101 Stories to Open the Heart and Rekindle the Spirit,* the authors, Jack Canfield and Mark Victor Hansen, commented:

> Hugging is healthy. It helps the body's immune system, it helps keep you healthier, it cures depression, it reduces stress, it induces sleep, it's invigorating, it's rejuvenating, it has no unpleasant side effect and hugging is nothing less than a miracle drug.
>
> Hugging is all natural. It is organic, naturally sweet, no pesticides, no preservatives, no artificial ingredients and 100 percent wholesome.[26]

Acupuncture

According to an article in *Cincinnati,* countless Americans undergo acupuncture from some 6,500 licensed practitioners. Complaints that have been helped by acupuncture include asthma, PMS, depression, arthritis, headaches, neck pain, stiffness and sexual dysfunction.

Writing in *Prevention* magazine, reporter Sharon Stocker said that Harvard, Yale, UCLA and other leading medical schools are all investigating acupuncture as a potential adjunctive treatment for a variety of health problems.

You can obtain more information from the American Academy of Medical Acupuncture (AAMA) at 1-800-521-AAMA. Or, you can write to the Academy at 5820 Wilshire Blvd., Suite 500, Los Angeles, CA 90036.

Acupressure Massage

In discussing acupressure and other types of Oriental bodywork in the book, *Alternative Medicine—The Definitive Guide,*[27] the authors said that whereas acupuncture uses needles, acupressure uses the pressure of the fingers and hands. Moreover, acupressure is the older of the two techniques. It has been found to be an effective technique that people can use to relieve the pain and discomfort of headaches and other disorders.

You can do acupressure either on yourself or it can be done by a friend. Your hands should be clean and your nails trimmed; the room should be warm and quiet. Pressure should be applied slowly with the tips or balls of the fingers and held at the point of discomfort for one to three minutes.

Strength Training

According to an article by Jane E. Brody in the January/February 1995 *Saturday Evening Post,* strength training is an important part of any physical fitness program. It can lower your total cholesterol, increase your good cholesterol and lessen your chances of developing heart disease. It also can reduce your risk of developing back problems and joint problems and give you more self-confidence.

"Strength training" is the process of building muscle power by lifting weights or working against resistance. Special equipment, like Nautilus or Universal machines, have been designed to help you accomplish this training. Yet, Brody says that most of the recommended exercises can be done with nothing fancier than a few full cans from your supermarket, or plastic bottles filled with water or sand.

UltraClear® and UltraClear® Sustain*

These powdered nutritional products are designed to provide nutritional support to patients with dietary needs related to he-

*Information about these products can be obtained from HealthComm International, Inc., 5800 Soundview Dr., Bldg. B, P.O. Box 1729, Gig Harbor, WA 98335, 800-843-9660.

patic detoxification, chronic fatigue syndrome, arthralgia and myalgia, food allergies, chemical sensitivities and other disorders.

According to Drs. Sidney Baker, Leo Galland and Nick Nonas, these products are useful in helping patients with gas, bloating, diarrhea, nausea and other symptoms, including some found in people with yeast-related health disorders.

Here's a testimonial from a woman who had been troubled by severe vulvovaginal problems and other chronic health disorders. Although she improved on antifungal medications and other therapies, many of her symptoms persisted. Then, because of tests that showed that she had a "leaky gut," her chiropractor prescribed a six week program of simple foods and Ultra Clear Sustain. She said, "It changed my life."

Touch

In an article in the July 1994 issue of *Health Confidential,* Tiffany Field, Ph.D., of the University of Miami School of Medicine, said:

> Human touch has remarkable restorative effects on the body. It reduces anxiety, stress and depression. It promotes relaxation and relieves pain. It fosters sound sleep, bolsters the immune system and enhances productivity. And—it feels great.

In discussing the biochemistry of touch, Field said that when you are deprived of touch, stress hormones are produced that can make you feel jittery and weaken your immune system. In discussing methods of touch, she said that family members can exchange massages.[28]

She also recommended massage therapists. To find a licensed massage therapist, check the Yellow Pages under "massage therapy schools," or contact the American Massage Therapy Association, 820 Davis St., Suite 100, Evanston, IL 60201, 708-864-0123.

Pure Water

During the 1980s and early 1990s, numerous reports appeared in newspapers, magazines and on TV describing contaminated public water supplies in many communities. Such contaminants include lead, harmful bacteria and parasites. In a 1993 cover story article published in *Health Confidential*, Richard P. Maas, Ph.D., associate professor of environmental studies, University of North Carolina, made several suggestions that would help you obtain less contaminated water:[29]

- Purge your water faucet for one minute in the morning before you drink. Since lead leeches into the water much more rapidly than previously thought, you'll need to purge it again if it's been more than a few minutes since the last time you drew water. To save time and water, keep a gallon pitcher of water from a purged tap in your refrigerator.
- Have your water tested. Unless you do this, you do not know whether you need to be purging the faucet or whether the purging is effective. Your local water utility may provide you with a free test kit. Such kits are also available from other sources, including:

Suburban Water Testing Lab
4600 Kutztown Rd.
Temple, PA 19560
800-433-6595 ($35)

Clean Water Lead Testing
29½ Page Ave.
Asheville, NC 28801
704-251-0518 ($17)

National Testing Laboratories
Wilson Mills Rd.
Cleveland, OH 44143
800-458-3330 ($35)

- You also can purchase a water purification system. There are several types, including cation-exchange filters, reverse osmosis filters and distillation units. Costs range from less than $200 to $400.
- Buy bottled spring water or distilled water.

Good Light

Fluorescent light may be contributing to your health problems, especially if you work in a windowless office or building and spend little time outdoors.

Research in a new field called photobiology, studies how light interacts with life. It also shows how ordinary fluorescent lighting may play a part in causing fatigue, depression and other health problems in adults and children. Natural sunlight—or skylight—even when the sun isn't shining, contains all the colors of the rainbow. Yet, most artificial lighting contains only a few colors.

How does natural light work and what are some of the good things it can do for you? It favorably influences your immune system, your hormones, the formation of Vitamin D, the absorption of calcium and phosphorus, your ability to pay attention and much, much more.

To receive the health benefits of light you should spend at least half an hour a day outside in natural light and obtain special broad spectrum lights for your home and office. These lights, which fit into standard fluorescent fixtures, can be purchased at garden supply stores, hardware stores, some lighting departments and health food stores.[30]

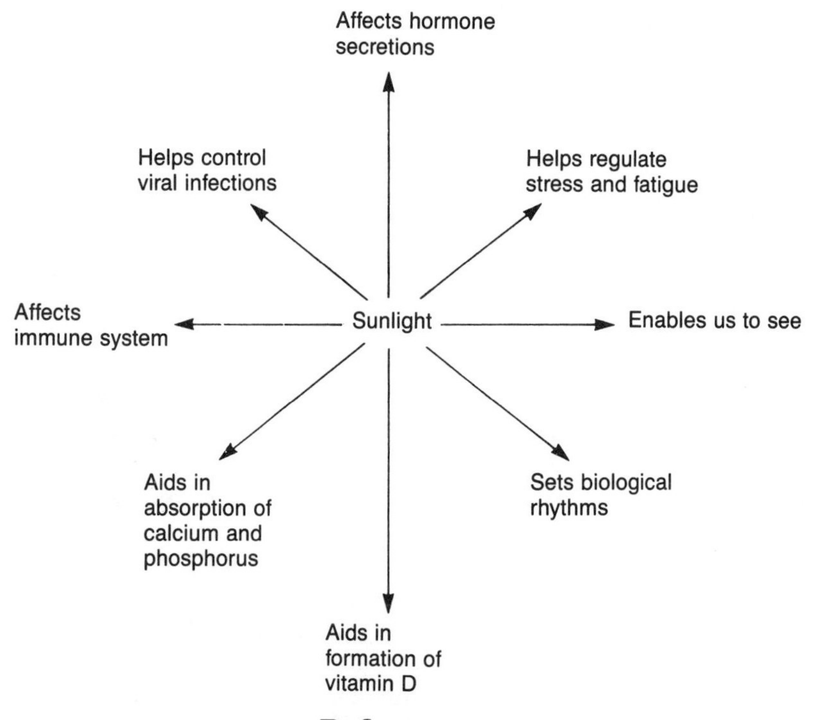

References

1. Barnes, B. O., "Basal Temperature vs. Basal Metabolism," *JAMA,* 1942; 119:1072–1074.

2. Barnes, B. O., with Dalton, L., *Hypothyroidism: The Unsuspected Illness,* Harper and Rowe, New York, 1976.

3. Crook, W. G., *The Yeast Connection,* Third Edition, Professional Books, Jackson, TN and Vintage Books, New York, 1986; p. 244.

4. Crook, W. G., *The Yeast Connection and the Woman,* Professional Books, Jackson, TN, 1995; pp. 547–558.

5. Jefferies, W. M., *Safe Uses of Cortisone,* Charles C. Thomas, Springfield, IL, 1981.

6. Jefferies, W. M., "Mild Adrenocortical Deficiency, Chronic Allergies, Autoimmune Disorders and the Chronic Fatigue Syndrome; A continuation of the Cortisone Story," *Medical Hypotheses,* 1994; 42:183–189.

7. Bliznakov, E. G. and Hunt, G. L., *The Miracle Nutrient, Coenzyme Q_{10}.* Bantam Books, New York, 1987; pp. 8, 64.

8. Frishman, W. H., "Coenzyme Q_{10}: A New Drug for Myocardial Ischemia?" *Medical Clinics of North America,* 1988; 72(1):243–258.

9. Whitaker, J., *Health and Healing,* Phillips Publishing Company, Inc., 7811 Montrose Rd., Potomac, MD 20854.

10. Gaby, A. R., in Dr. Jonathan V. Wright's *Nutrition in Healing,* Vol. II, Issue 11, November, 1995, P.O. Box 84909, Phoenix, AZ 85071.

11. Lockwood, K., et al, "Progress on Therapy of Breast Cancer with Vitamin Q_{10} and the Regression of Metastases," *Biochem. Biophys. Res. Commun.,* 1995; 212:172–177.

12. Klenner, F. R., "Virus pneumonia and its treatment with vitamin C," *J. South. Med. and Surg.,* 1948; 110:60–63.

13. Klenner, F. R., "The treatment of poliomyelitis and other virus diseases with vitamin C," *J. South. Med. and Surg.,* 1949; 111:210–214.

14. Klenner, F. R., "Observations on the dose and administration of ascorbic acid when employed beyond the range of a vitamin in human pathology," *J. App. Nutr.,* 1971; 23:61–88.

15. Stone, I., *The Healing Factor: Vitamin C Against Disease,* Grosset and Dunlap, New York, 1972.

16. Cathcart, R. F., "Vitamin C: titrating to bowel-tolerance, anascorbemia, and acute induced scurvy," *Medical Hypotheses,* 1981; 7:1359–1376.

17. Cathcart, R. F., "The third face of vitamin C," *J. of Orthomol. Med.* 1993; 7(4):197–200.

18. Newbold, H. L., *Dr. Newbold's Type A/Type B Weight Loss Book,* Keats Publishing Company, New Canaan, CT, 1991.

19. Lindenbaum, J., et al, "Neuropsychiatric Disorders Caused By Cobalamin (Vitamin B_{12}) Deficiency in the Absence of Anaemia or Macrocytosis," *N. Engl. J. Med.,* 1988; 318:1720–1728.

20. Editorial, "Pharmaceuticals from plants: great potential, few funds," *The Lancet,* 1994; 343:1513–1515.

21. Schwitters, B., *OPC in Practice—bioflavanols and their application,* Alfa Omega Publishers, Rome, 1993; pp. 6–7.

22. Passwater, R. A., Kandaswami, C., *Pycnogenol, the Super "Protector" Nutrient,* Keats Publishing Co., New Canaan, CT, 1994; pp. 1, 108.

23. Murray, M. T., *The Healing Power of Herbs,* 2nd Ed., Prima Publishing, Rockland, CA, 1995.

24. Baker, S. M., personal communication, May 1985.

25. Galland, L., in Crook, W. G., *The Yeast Connection,* Third Edition, Professional Books, Jackson, TN and Vintage Books, New York, 1986; pp. 364–366.

26. Canfield, J. and Hansen, M. V., *Chicken Soup for the Soul,* Health Communications, Inc., Deerfield Beach, FL, 1993, pp. 16–18.

27. Compiled by The Burton Goldberg Group, *Alternative Medicine—The Definitive Guide,* Future Medicine Publishing, Inc., Puyallup, WA 1994; p. 110.

28. Field, T., *Health Confidential,* July 1994, 55 Railroad Avenue, Greenwich, CT 06830.

29. Maas, R. P., *Health Confidential,* 55 Railroad Ave., Greenwich, CT 06830, December 1993.

30. Cerda, G. M. and Parry, B. L., "The Effects of Bright Light Therapy on Symptoms of Depression, Anxiety, and Hibernation in Patients with Premenstrual Syndrome," *Journal of Women's Health,* 1994; 3(1):5–15.

CHAPTER · TWELVE

Other Topics of Interest

Sweeteners

Every week I receive phone calls and letters from people who say, "If I can't use sugar, what can I use?" Sometimes I say something like this, "Everybody likes sweets, certainly I do. Even young infants will smile if you put something sweet in their mouths. Yet, during the early stages of your diet, avoiding sugar, corn syrup, honey, molasses, maple syrup and other simple carbohydrates is an essential part of any candida treatment program. But diets are not forever and after a few weeks (or months) you can experiment."

I also tell people that if they tolerate fruits they can use them in some of their recipes. Or, they can put a banana, peach or other sweet fruit in a blender with a little water and make a "slurry" to pour on cereal or add to other foods.

But, I also have some good news about two natural sweeteners, fructooligosaccharides (FOS) and stevia, and one of the artificial sweeteners, pure liquid saccharin.

FOS: A Healthful Sweetener*

The October 1982 issue of *Let's Live* magazine had an article entitled, "FOS: A Healthful Sweetener (No kidding!)," that explained the sweetener. In it, Betty Kamen, Ph.D., said:

*You'll find more information about FOS in Robert Crayhon's 48-page booklet, *Health Benefits of FOS,* Keats Publishing Co., New Canaan, CT, 1995.

FOS are sucrose molecules to which 1, 2 or 3 additional fructose molecules have been linked in sequence. FOS are widely distributed in a variety of edible plants, such as vegetables and grains and some fruits. They're not absorbed by human digestive enzymes in the gastrointestinal tract or pancreas, but are utilized by your friendly intestinal bacteria.

FOS promotes the growth of beneficial bacteria, including bifidus and lactobacilli. And in doing so, they interfere with the growth of disease-causing organisms.[1]

To get further information about FOS, I wrote to Beatrice Trum Hunter, food editor of *Consumers' Research,* who has written and published more than a dozen books on food issues. In early February 1994, she sent me an article in a peer-reviewed journal, which documented the safety and benefits of FOS. Here is an excerpt from this comprehensive, scientific article:

> Fructooligosaccharides can have beneficial effects as food ingredients . . . In Japan, FOS are considered food, not food ingredients, and are found in more than 500 food products, resulting in significant daily consumption . . . Numerous in-vitro and in-vivo studies have been conducted to evaluate the potential toxicity of FOS to animals and man . . . *Results provided no evidence that FOS possessed any genotoxic potential* (emphasis added).[2]

I received more information about FOS from Marjorie Hurt Jones, R.N., and Nicolette M. Dumke. Each of them have coauthored several books that deal with allergy cooking. In discussing FOS, Marge pointed out one downside: cost. And she said:

> With supplies in this country still quite limited, supplement companies have jumped in and are selling 60–100 grams of FOS for $10–$12. (100 grams equals ½ cup.) If you want to use 1–2 teaspoons a day, therapeutically, that may be okay. But if you want to bake cookies—you will need $20–$40 worth of FOS for one batch! Ouch![3]

You can find pure FOS in powder form or in capsules. And some producers combine FOS with a probiotic. Taking these products in powder or capsule form can help you develop a healthy balance of bacteria in your intestinal tract.

Another source of FOS that I learned about recently is Jerusalem artichoke powder. It can be sprinkled on cold or hot cereal as a topping or used in a bread, muffin or cookie recipe as a partial flour replacement (up to 10 percent). More information about this product can be obtained from HealthCo International, P.O. Box 544, Bloomingdale, IL 60108 (800-477-3949).

Stevia

This herbal product, which comes from a perennial shrub of the aster family, is estimated to be 150 to 400 times sweeter than sugar and contains no calories. It has been used for centuries in parts of South America and consumed by millions of people in other countries, including China and Japan.

According to an article in the February 1996 issue of *New Age Journal,* Japanese researchers have studied stevia extract and found it to be without health risk. Moreover, they combine it in numerous food products, including candies, ice cream and soft drinks.

Yet, in spite of its wide-spread use, the FDA allows it only to be used as a "dietary supplement." It does not allow companies that produce and market it to recommend it as a sweetener.

Where can you get stevia products? You may be able to find them in your health food store. You also can obtain more information from several companies who market and distribute it. Here are their numbers: 417-753-3999, 800-947-6417, 518-239-8327 and 800-478-3842.

Based on information I've received from various sources, including Mark Blumenthal, executive director of the American Botanical Council,[4] stevia does not encourage the growth of yeast. I certainly feel it's worth a try.

Saccharin

U.S. laws require that products containing saccharin carry this label: Use of this product may be hazardous to your health. This product contains saccharin, which has been determined to cause cancer in laboratory animals.

Notwithstanding, several recent reports of peer-reviewed literature point out that saccharin is safe. Here's an excerpt of an article published by researchers from South Dakota State University:

> Almost from its discovery in 1879, the use of saccharin as an artificial, non-nutritive sweetener has been the center of several controversies regarding potential toxic effects, most recently focusing on the urinary bladder carcinogenicity of sodium saccharin in rats when fed at high doses in two-generation studies.
>
> No carcinogenic effect has been observed in mice, hamsters or monkeys, and numerous epidemiological studies provide no clear or consistent evidence to support the assertion that sodium saccharin increases the risk of bladder cancer in the human population.[5]

Clear liquid saccharin is available in most any drugstore or grocery store. Brand names include Fasweet (Schering-Plough) and Sweeta (Squibb). I personally use small amounts of saccharin to sweeten occasional cups of coffee and tea. And I do not worry about increasing my risk of bladder cancer.

Aspartame

Aspartame (NutraSweet) is consumed by more than 100 million people in the United States. It has been endorsed by the FDA and a number of medical organizations. Nevertheless, reports I've received from professionals and nonprofessionals suggest that aspartame causes many adverse reactions, including headaches, possible cramps, depression and other symptoms.

If you would like more information, write to H. J. Roberts, M.D., 300 27th St., West Palm Beach, Florida 33407-5299. This board-certified internist has published two books on dealing with aspartame: *Sweet'ner Dearest: Bittersweet Vignettes About Aspartame (NutraSweet)*[6] and *ASPARTAME (NutraSweet), Is It Safe?*[7]

Alternative Medicine

According to a study published in the January 28, 1993, issue of the *New England Journal of Medicine,* one-third of all Americans have sought help from alternative practitioners.[8] And most of the time they haven't told their own physicians. Their conviction that such therapies help is documented by the money they've spent: $13.7 billion. Moreover, they've paid $10.3 billion themselves, since their private health insurance and Medicare or Medicaid doesn't cover it.

In the introduction to his 1,068-page book, *Alternative Medicine: The Definitive Guide*, editor Burton Goldberg said:

> Two systems of health care are available in this country today: conventional Western medicine and alternative medicine. The first is the world of the American Medical Association; medical doctors who practice by the book . . . Conventional medicine is superb when it comes to surgery, emergency, and trauma.
>
> But there's no question that alternative medicine works better for just about everything else, especially for chronic degenerative diseases like cancer, heart disease, rheumatoid arthritis and for more common ailments, such as asthma, gastrointestinal disorders, headaches and sinusitis . . .
>
> Alternative medicine is more cost effective over the long term. Because it emphasizes prevention and goes after causes rather than symptoms, it doesn't trap people on the merry-go-round that begins with one drug and ends up requiring them to take others to compensate for the side effects each one causes. Many alternative methods work by assisting your own body to heal itself instead of introducing strong drugs.[9]

Goldberg pointed out that it takes time for new ideas to be accepted. He also cited the story of Ignaz Semmelweiss, the Austrian physician who was ridiculed and fired from the staff of a Vienna hospital. His offense: He asked his colleagues to wash their hands after working on cadavers before carrying out pelvic examinations! Yet, it took doctors 30 years to "catch on."

I found further information and support for alternative medicine in *USA Weekend*. In the sixth annual special health issue entitled, *Can Alternative Medicine Help My Headache*, many different alternative therapies were listed and discussed for treating five common ailments: headaches, arthritis, back pain, extra pounds and asthma.

The article cited the ideas and approaches of many experts, including James Gordon, M.D., of Georgetown University, Dean

Ornish, M.D., and Anthony Rosner, Ph.D., research director of the Foundation for Chiropractic Education and Research. This issue, which was devoted entirely to alternative medicine, cited and quoted many people, including professionals and nonprofessionals. Included also were ways "not to get snookered." Some of these were: See a physician for acute conditions; look for credentials; ask for referrals; be skeptical of unrealistic claims and have realistic expectations.

In early 1996, alternative medicine received further attention and support. Here are a few highlights.

- The King County (Washington) council voted to establish the nations first government subsidized natural medicine clinic. According to a Saturday, February 25, 1995 article in the *Seattle Post Intelligencer,* the clinic will offer medical treatment using herbal remedies, manipulative therapies, acupuncture and nutrition. The staff of the clinic will include a medical doctor, as well as a doctor of naturopathy, a researcher and a nutritionist. The clinic will be managed by Bastyr University, a Seattle based natural medicine institution.
- Joseph Pizzorno, Jr., N.D., President of Bastyr University, participated in a national radio debate with the author of a piece published in the Opinion/Editorial Section of the *New York Times.*
- A number of major medical schools including Harvard, Columbia and Stanford, conducted programs on alternative medicine between 1996 and 1999.

Herbal Remedies*

Herbal remedies aren't all that new. What's more, during my days in medical school I learned about digitalis, the main heart medicine derived from the foxglove plant. Here are a few others that I know about, including some that I learned about many years ago and others I first heard about during the past decade:

*For more about herbal remedies and alternative medicine, see pages 260–262.

Herbal Teas, Lime Juice and Vitamin C

Indians in Canada discovered that tea made from the leaves and bark of the *Arbor vitae* tree helped many members of their tribe with various complaints. In the 17th Century, a friendly Indian passed this information to French explorer Jacques Cartier, whose soldiers in Quebec were dying of scurvy. When the soldiers drank the tea, their scurvy went away and many lives were saved. Some 75 years later, Dr. James Lind used lime juice to prevent scurvy in sailors on British ships.

Ginkgo Biloba

This "folk remedy" from the Ginkgo tree has been used in China for a long, long time. A recent scientific study provides support for this therapy. Here are excerpts from a 1992 article in *The Lancet*:

> Extracts from the leaves of Ginkgo biloba . . . have been used therapeutically for centuries. Ginkgo, like ginseng, is mentioned in the traditional Chinese pharmacopoeia. . . . In Germany and France, such extracts are among the most commonly prescribed drugs.

The article reviewed 40 published controlled trials, most of which were conducted in Germany and France. They found that eight trials were of good quality. They especially noted that elderly people with difficulties of concentration and memory, absentmindedness, confusion, lack of energy, depressive mood, anxiety, dizziness, tinnitus and headache were improved. And patients described in two of the trials showed "an increase in walking distance."

> In most trials, 120 to 160 mg. a day, divided into three doses, was used . . . Treatment must be four to six weeks before positive effects can be expected. Whether beneficial effects remain when treatment is stopped is unknown. The only drawback worth mentioning is the high cost . . .[10]

Bromelain

This plant substance, derived from the stem of the pineapple plant, contains enzymes that help in the treatment of many conditions. According to Dr. Michael Murray, it was introduced as a therapeutic agent in 1957 and, since that time, more than 200 scientific papers on the therapeutic applications have appeared in the medical literature.

Bromelain has been used in the treatment of arthritis, athletic injuries, bronchitis, painful menstruation, digestive problems, sinusitis and many others. Usual dose: 250-500 milligrams three times a day between meals.

Echinacea

According to Earl Mindell,[11] echinacea comes from the purple coneflower and was used by native Americans to treat snake bite, fevers and old, stubborn wounds. In his discussion, Mindell pointed out that there has been a renewed interest in echinacea in the United States because of its positive effect on the immune system. He also noted that it has been used successfully in treating candida, psoriasis and eczema.

Other Herbal Remedies

Mindell also discussed golden seal, astragalus, garlic, ginger, ginkgo, ginseng, saw palmetto, feverfew and Pau d'Arco. This latter herbal remedy comes from the bark of two South American trees. During the past decade, I've received reports from both professionals and nonprofessionals telling me of the value of this herb in treating people with yeast-related health problems.

Parasites

A sugar-free special diet, antifungal medications, avoidance of indoor chemical pollutants and other treatment measures help many people with chronic health disorders get well. But some people continue to experience problems. If that is your situation, intestinal parasites may be playing a part in making you sick.

Several years ago at a candida conference, Dr. Leo Galland reported that many patients with yeast-related health problems also are infected with parasites.

In 1989, Dr. Galland and a colleague, Herman Bueno, conducted a two-year retrospective study of 218 patients who came to his clinic complaining of chronic fatigue. *Giardia lamblia* infection was identified by rectal swab in 61 of these patients. In discussing their observations, they said:

> Cure of giardiasis resulted in clearing of fatigue and related "viral" symptoms (myalgia sweats, flu-like feeling) in 70 percent, fatigue in 18 percent and was no benefit in 12 percent. This study shows that giardiasis can be present with fatigue as the major manifestation, accompanied by minor gastrointestinal complaints, and sometimes by myalgia (muscle aching).[12]

Another parasite, *Blastocystis hominis,* also appears to be a troublemaker. According to an article in *Practical Gastroenterology,* by Martin J. Lee, Ph.D.:

> *Blastocystis hominis* has greater prevalence than any other parasite, but often goes undetected because of poor laboratory technique. At Great Smokies Diagnostic Laboratory, *Blastocystis* is found in 15 percent or more of clinical specimens. The weight of evidence supports treating it as a potential pathogen.[13]

You'll find much more information about parasites in Anna Louise Gittleman's 1993 book, *Guess What Came to Dinner—Parasites and Your Health.*[14] This comprehensive but easy-to-read and understand book provides readers with just about everything they need to know about parasites, including symptoms, diagnosis, treatment and prevention. She also lists a number of laboratories that specialize in parasite testing, including:

- Great Smokies Diagnostic Laboratory, 63 Zillicoa Street, Asheville, NC 28801 (800-522-4762).
- Lexington Professional Center, 133 E. 73rd St., New York, NY 10021. (212-988-4800).
- Meridian Valley Clinical Laboratory, 24030 132nd Ave., S.E., Kent, WA 98042 (800-234-6825).
- Parasitic Disease Consultants Laboratory, P.O. Box 616, 2177-J Flintstone Dr., Tucker, GA 30084 (770-496-1370).
- Parasitology Laboratory of Washington, Inc., 2141 K St., N.W., Suite 408, Washington, DC 20037 (202-331-0287).

Mercury/Amalgam Dental Fillings

The hazards of these fillings have been talked about by a handful of dentists during the past decade. According to these dentists, these fillings may cause toxic reactions and play a part in making people sick. Symptoms include fatigue, headache, central nervous system dysfunction, muscle and joint pains and disturbances in other parts of the body.

An opposite point of view is expressed by the American Dental Association, which represents at least 75 percent of the nation's dentists. This organization continues to support the use of amalgam as a safe, effective method of tooth restoration. This controversy was brought into the public "mainstream" on CBS's "60 Minutes," December 15, 1990.

On this program, Dr. Murray J. Vimy of the University of Calgary Medical School and Alfred Zamm, M.D., of Kingston, New York, presented information on the potential and actual toxic effects of mercury/amalgam fillings. Vimy had placed silver fillings in the mouths of pregnant sheep. Three days later, mercury was found in the blood of mothers and fetuses and in the amniotic fluid. Two weeks later, it was found in other tissue samples.

Along with their comments and presentations of scientific data were testimonials from patients whose fatigue and other chronic health problems were "turned around" following the removal of mercury/amalgam fillings.

Based on information I received from Dr. Zamm and other professionals and patients, I'm convinced that silver or mercury dental fillings play a part in making many people sick. So do "root canals." According to Dr. Zamm, substances that are inserted into the teeth during this procedure are cytotoxic and can leak into the circulation, causing a suppression of the immune system.

Further information about the toxic effects of amalgam dental fillings and root canals can be obtained from:

- The Environmental Dental Association (EDA) 800-388-8124.
- *Holistic Dental Digest* by Jerry Mittleman, DDS. This bimonthly newsletter is published by The Once Daily, Inc. Sub-

scription rate is $9.50/year. For a sample copy, send a long, self-addressed, stamped envelope to Dr. Mittelman, 263 West End, #2A, New York, NY 10023.

Books that provide information on mercury/amalgam dental fillings include:

- *The Complete Guide to Mercury Toxicity from Dental Fillings,* Joyal Taylor, DDS, EDA Publishing, 9974 Scripps Ranch Blvd., Suite 36, San Diego, CA 92131. $14.95 plus $3.00 shipping.
- *Are Your Dental Fillings Poisoning You?* Guy, S. Fasciana, DMD, Keats Publishing Co., 27 Pine St., New Canaan, CT 06840.
- *The Toxic Time Bomb,* Sam Ziff, DDS, BioProbe, Inc., P.O. Box 6080, Orlando, FL 32860.

Free Radicals and Antioxidants

About 10 years ago I began hearing about free radicals and antioxidants. Although I read about them in medical and lay journals, I must confess that I still had more questions than answers. Then at a health food convention in 1992, I heard a presentation dealing with these topics, by David J. Lin, B.S. In his lecture and in a subsequent book, Lin discussed these topics in an easy-to-understand format. In the introduction to his book, he said:

> Uncontrolled free radical processes may be critically involved in the cause and progression of numerous disease conditions, conditions which before seemed unrelated.

He also noted that simple nutrients, called antioxidants, are perhaps the best natural defenses against these harmful substances. Lin said:

> *Free radicals are molecules with electrons which are unpaired* [emphasis added]. Stable molecules have electrons in pairs, like a buddy system. But if a molecule has an electron which does not have a partner, it becomes unstable and reactive . . . *a free radical* [emphasis added]. It will steal an electron from a stable molecule.

Once the stable molecule loses an electron, it becomes another free radical. This second free radical will steal an electron from a third molecule, and a destructive cycle begins. Each time a molecule loses an electron it is damaged and will damage another molecule.

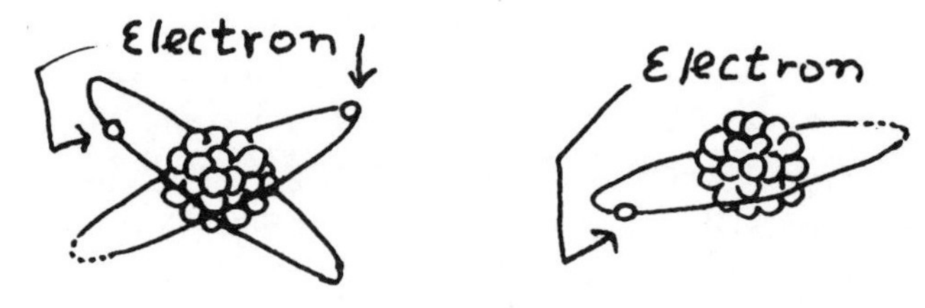

STABLE MOLECULE UNSTABLE MOLECULE

Where Free Radicals Come From

Free radicals come from several sources. Here are a few of them:

- Our bodies are constantly making them. They help our immune system destroy bacteria and viruses. Therefore, free radicals are important to health. They aren't "evil." *It is only when they are produced in excessive amounts that they damage the body.*
- Tobacco smoke, pesticides, herbicides and other air pollutants also encourage the production of free radicals.
- Chain reactions. According to Lin, free radicals produce a "domino effect." One free radical produces a second, which in turn produces a third, and so on. And this is what makes free radicals so dangerous.

In his continuing discussion, Lin compares the effect of free radicals to what happens on a freeway when a tire rolls over a big sharp nail. POW! The tire blows. The car swerves out of control and crashes into another. Then another car piles on. Then another and another, until you have a tremendous pile up. The cars represent molecules in our cells and the nail is a free radical.

Your body is composed of billions of healthy cells of all sorts; cells arranged in an orderly fashion; cells with normal membranes and layers of nuclei. Free radicals damage these cells in various ways, allowing bacteria and viruses to enter.

According to Lin:

> The cells damaged by free radicals can accumulate to become full blown disease states. Just as one tiny nail on the freeway can cause a major catastrophe, so a free radical can cause and worsen serious disease conditions, including heart disease, cancer, cataracts, arthritis, Parkinson's disease and many others.

Antioxidants

Before discussing antioxidants, let's talk about "oxidants." This term comes from the word oxygen. Quoting Lin again,

> Leave oils and meats outside too long and they become rancid. Rancidity involves oxidants in a process called oxidation, which is the reaction of oxygen with fatty acids and proteins to form free radicals, eventually causing spoiling.

Antioxidants fight oxidation by neutralizing free radicals which cause it. Using the analogy of the nail on the freeway, Lin said that antioxidants function as highway sweepers to get rid of the nails and avoid damaging wrecks.

In coping with free radicals, we can make appropriate changes in our lifestyle. We can stop smoking and stop polluting our homes with toxic chemicals. We can eat foods rich in the antioxidant nutrients, especially fruits, vegetables, grains and nuts. These nutrients function together as a team to quench free radicals.

We also can take nutritional supplements, especially the four antioxidant nutrients—beta carotene, vitamin C, vitamin E and the trace mineral selenium. In concluding this discussion, Lin said that there is a growing body of scientific evidence that substantiates the effectiveness of such therapies. Yet, he said,

> Antioxidants are vital defenders against excessive free radical attacks. The role of proper nutrition in preventing disease is becoming clearer . . . Once dismissed by medical authorities as minor players in maintaining health, vitamins and minerals are beginning to be used to treat a wide variety of serious ailments.

As the role of antioxidants in combating free radical pathology becomes clearer, we will hopefully be able to effectively treat diseases which have plagued us for ages.[15]

Natural Progesterone

In a 1993 book entitled, *Unmasking PMS—The complete medical treatment plan,*[16] Joseph Martorano and his co-authors recommended dietary changes and nystatin for women with a variety of problems, including PMS, fatigue, depression, food sensitivities, inability to concentrate and other symptoms.

They also included an eight-page chapter on natural progesterone. In their discussion, they said that natural progesterone, extracted from yams or soybeans, is chemically identical to progesterone produced naturally by the body.

These professionals stated that oral micronized tablets of progesterone, which first became available in the late 1980s, were a "giant step forward" in treating many of their patients with PMS. Additional support for this form of natural progesterone was published by Vanderbilt University Medical Center researchers in the *American Journal of Obstetrics and Gynecology.* Here are excerpts from an abstract of their study.

> The oral route of progesterone administration has long been considered impractical because of poor absorption and short biologic half-life. Recent reports suggest that micronization of progesterone enhances absorption and increases serum and tissue levels of progesterone.[17]

I found more information about this hormone in Dr. John R. Lee's fascinating book, *Natural Progesterone: The Multiple Roles of a Remarkable Hormone.*[18]

In discussing natural hormones with Jeffrey Bland, Ph.D.,[19] Lee emphasized the differences between natural progesterone and the progestogens. And he said that the latter hormones, such as Provera, while heavily advertised, can and do cause many side effects—and that such side effects are *not* caused by natural progesterone.

Lee noted that natural progesterone lessens premenstrual irritability, water retention, depression and loss of libido. Yet, he said that it isn't a "quick fix" and that it is important for it to be used in a way that matches the normal menstrual cycle. He also said that if it's not used at an appropriate time of the cycle, women will develop irregular bleeding.

Natural Progesterone and the Prevention of Osteoporosis

In a book, *Hormone Replacement Therapy, Yes or No?*[20] Betty Kamen, Ph.D., in a chapter on natural progesterone reviewed the observations of Dr. Lee. She also cites the scientific reports of Jerilynn Prior, M.D.,[21,22] of the Endocrinology and Metabolism Division of the University of British Columbia, who found that natural progesterone helped prevent and treat osteoporosis.

In her research study, Prior gave progesterone for ten days a month, plus one gram of calcium, to 61 women with menstrual cycle disturbances. As a result, the bone density in these women increased by two percent. By contrast, 14 women who were given placebo (blank) pills along with the calcium, lost two percent of their bone density during the year of the study.

Comments by Health Professionals

To learn more about hormone therapy in women, I talked to a number of health professionals, including Dr. C. Orian Truss. He told me that hormone dysfunction will often disappear following anticandida therapy. Moreover, in his book, *The Missing Diagnosis,* he said:

> By unknown mechanisms, progesterone greatly aggravates yeast growth in women . . . Hormone administration not only fails to correct the problem, but indeed may aggravate it.[23]

However, other health professionals, including Jean Rowe, R.N. of Denver, told me that they had found micronized progesterone especially helpful in women with PMS. The usual dosage ranged from 50 to 300 milligrams, twice daily.

My Comments

I realize that I have barely scratched the surface in discussing progesterone therapy. Yet, from what I've read and heard during the past two years, supplemental natural progesterone seems to help many women with PMS and/or menopausal symptoms. It also may play an important role in preventing osteoporosis.

Obviously, natural progesterone should be only one part of a program, which includes a good diet, exercise and nutritional supplements. Finally, it must be prescribed and monitored by a physician.

Yeast-Related Drunkenness

In *The Yeast Connection*[24] I told the story of Charlie Swaart, an Arizona man, who for more than a decade would become a sloppy, overbearing, hostile and sometimes even violent drunk, even though he hadn't consumed any alcohol. Moreover, on at least one occasion, Charlie was picked up for drunk driving.

Happily for Charlie, his Arizona physician learned of a report from Japan that described drunkenness caused by an overgrowth of *Candida albicans* in the digestive tract. So his physician placed him on nystatin and he showed significant—even dramatic—improvement.

A Japanese Study of Yeast-Related Drunkenness

In a 1970s article, "A Review of the Literature on Drunken Symptoms Due to Yeast in the Gastrointestinal Tract," Dr. Kazuo Iwata listed 24 reports of patients who showed "drunken syndrome" following ingestion of ordinary foods. Seventeen of the patients were males and seven were females. Their ages ranged between 16 months and 75 years; most were middle-aged. *Candida albicans* was isolated in the gastric and duodenal juices of 14 of the patients and other yeasts were found in the remainder. A number of the patients were troubled by a partial blocking of the intestinal tract, which slowed the passage of the contents through the bowel.

In discussing the differential diagnosis, Iwata said:

> The differential diagnosis is essentially that of alcoholism. Patients' chief complaint is suffering occasional or frequent symptoms of acute intoxication, despite not having taken a drink of any intoxicating beverage, but they're apt to be mistrusted by

doctors, especially at the initial inspection of the patients who have breath with an alcoholic odor.[25]

Stories of People with Yeast-Related Drunkenness

During the past decade I've received several dozen calls and reports from people with yeast-related problems who reported feelings of drunkenness. Here are the stories of some of them.

Albert

This 26-year-old computer consultant and lease broker called me and told me he had been troubled by yeast-related intoxication. His problems developed after he had taken repeated courses of broad-spectrum antibiotics. His drunk-like feelings caused him to engage in all sorts of bizarre behavior. Then he went to Dr. Carol Englender, who put him on a yeast-free diet and nystatin. When I last talked to him he said, "I feel great."

Bradley Allen Gray

> I'm a 47-year-old former journalist suffering from candida-related health problems for nearly 27 years as I now recognize. My severe attacks began about six years ago. Since then I've been convicted of drunk driving and registered with medical documentation blood alcohol levels like .398, .40, .289, etc. Each time I was totally lucid and cognizant, and not drinking.
>
> I got tried by a jury, which deliberated 11 minutes. They sent me to jail for nine months. I was laughed at because of my complaints. Even in jail, the physician threatened to lock me up if I persisted with this "crap about a distillery in your gut." I had four severe attacks while in jail.

Melinda

This 45-year-old woman was arrested for drunk driving. One month before her arrest, Melinda suffered a compound fracture of both bones in her right leg. A day or two after her injury, she developed a severe infection; antibiotics were prescribed, which she took for the ensuing three weeks.

Three hours before her arrest, Melinda went to dinner with a friend. During this time she consumed two large glasses of a sugar-containing cola beverage and two 8 oz. glasses of wine. While driving home, she was stopped by a police officer. He noticed alcohol on her breath, took her to the police station and drew a sample of her blood. Analysis by a toxicologist showed her blood alcohol level of .205.

Because I felt that Melinda's elevated blood alcohol level was yeast-related, I flew to Wisconsin and testified in her behalf. Yet, my efforts and those of her lawyer, did not meet with success. In a news article headlined, "Woman Found Guilty of Driving Drunk," the reporter said, "Although a jury didn't completely buy the 'auto-brewery syndrome' defense in a drunken driving case, it didn't discount the theory . . ."

Melinda's case attracted the attention of lawyers throughout the United States. In an article in *The Champion*, published by the National Association of Criminal Defense Lawyers, entitled "A New Defense for DUI Cases: The Auto-Intoxication Syndrome," W. J. Edwards said,

> Medical evidence should now be considered by the DUI practitioner. Those defending a DUI case should look more carefully at the medical history of their clients to determine if such yeast fungus could exist inside the stomach . . . Despite the jury's verdict in Wisconsin, the author believes this defense is viable. Other jurors may be persuaded to accept the effect that the common yeast has upon the human body.[26]

I've received many other calls and letters from people who say, "At times I feel spaced out, not-with-it, drowsy and dazed—almost like a person who's had too much to drink—especially after I've cheated on my diet."

Immunotherapy

Immunotherapy stimulates or strengthens the immune system by using various kinds of extracts or vaccines. For example, you may have received vaccines, which help strengthen your immunity to diphtheria, tetanus, measles, polio and other diseases. Immunotherapy was first reported to help patients with hay fever in 1911. Since that time, it has been used in treating patients bothered by allergic rhinitis (hay fever) and asthma caused by pollens, dusts and molds.[27]

For the past 30 years, immunotherapy has been used by some physicians in treating patients with other types of sensitivities, including foods, hormones and yeasts.

In a 1962 report in the *Annals of Allergy,*[28] an Israeli physician, A. Liebeskind, described improvement in patients with candida-related health problems following the use of candida extracts. Symptoms in these patients included vulvitis, bronchitis, migraine and gastrointestinal complaints.

In the early 70s, several reports appeared in the medical literature describing the effectiveness of candida immunotherapy in patients with yeast vaginitis. One of these reports also reminded gynecologists and allergists that allergy may involve the vulvovaginal tract.[29] The reporter pointed out that specific anti-allergic therapy may be indicated in many resistant cases of chronic vaginitis.

Observations of C. Orian Truss, M.D.

In his paper, "Restoration of Immunologic Competence to *Candida albicans,*" C. Orian Truss, M.D., discussed diet, antifungal medications, the avoidance of antibiotics and other treatment measures. Included in the treatment program was the use of extracts of *Candida albicans*. Here are excerpts from his report.

> Injection of an extract of *Candida albicans* constitutes an attempt to strengthen the immune response to the yeast in a patient who is actively infected by the organism . . . Marked variation is exhibited among patients in the pattern of their allergic responses. It is not surprising, therefore, that in this unique situation it has proved impossible to standardize a program of injection therapy.
>
> The strength of the extract and the interval between injections varies not only among patients, but also in the same patient at different times of recovery.[30]

In his continuing discussion, Dr. Truss stated that he had found it "unrewarding" to steadily increase the strength of the allergy extract—a technique or method that is often successful with conventional injection therapy for inhalant allergy. He said in his experience, the use of much weaker doses in the range of 10^{-5} to 10^{-15}, were usually used in his practice and found to be effective. Yet, he emphasized that patience, time and careful re-evaluation of the patient is essential for optimal results.

Observations of Joseph B. Miller, M.D.

During the past two decades, Joseph B. Miller, M.D., has been the leading researcher and clinician in studying still another type of immunotherapy* in treating patients with food and inhalant allergies and other illnesses. Moreover, he's published his observations on the use of immunotherapy in helping people with PMS and other progesterone-related symptoms, and herpes and other viral infections.[31] During the past decade, Dr. Miller has found that using a similar technique, he has been able to help many patients with vaginitis and other candida-related disorders.

Enzyme Potentiated Desensitization

This type of immunotherapy uses extremely small doses of allergens in an attempt to desensitize the patient to his allergies. The doses used are the same or smaller than those employed in diagnostic skin prick testing. This method of therapy was pioneered by a British physician in 1966 who found that a single dose of grass pollen with beta glucuronidase could be as effective as a long course of conventional desensitizing injections.[32]

At this time, enzyme potentiated desensitization therapy is being used by fewer than 50 physicians in the United States. Further information, including an instruction booklet, can be obtained by writing to the International EPD Society, c/o W. A. Shrader, Jr., M.D., 141 Paseo de Peralta, Santa Fe, NM 87501. (Please enclose a check for $5 to cover postage/ handling and cost of the booklet.)

---

*Information about the methods of testing can be obtained from Dr. Miller's publications or by writing him at the Miller Center for Allergy, 5901 Airport Blvd., Mobile, AL 36608.

AIDS

A study published in the *New England Journal of Medicine* more than a decade ago showed that the majority of AIDS victims developed candida infections in the mouth—just as do other patients with weakened immune systems. And during the past decade, physicians studying and treating people with this distressing disease have found that anti-yeast therapy helps many of them.

To obtain more information on the subject, I interviewed Drs. Alan Levin and Carol Jessop of San Francisco and Dr. Joan Priestley of Anchorage, Alaska. Here's a consensus of their observations:

- Because of their weakened immune systems, all people with AIDS are troubled by *Candida albicans* infections. Accordingly, antifungal therapy is an important part of their treatment.
- People who take azole drugs for many months should be carefully followed by their health providers. Individuals who are taking Nizoral should receive blood tests to monitor liver function every two weeks; those taking Diflucan or Sporanox should receive the blood tests every four to six weeks.
- As with other yeast-related disorders, diet, nutritional supplements and other measures are also important. Included among these are the use of nonprescription agents, including garlic and acidophilus.
- Individuals with AIDS, as well as those who are HIV positive, should consume a nutritious low sugar diet, take nutritional supplements and avoid environmental pollutants.

More About Diagnostic Studies

How does a physician make a diagnosis? How does she/he find out what's making you "sick"? There are three main methods.

1. Your *history*
2. Your *physical examination*—what the physician can determine with his/her eyes, ears, and touch

3. *Tests,* including blood tests, urine tests, stool examinations, X-rays, and other more complex laboratory studies.

Your History or Story

Let's talk about your history first. Many physicians feel that a patient's story provides more information than any other diagnostic method. Here are typical things your physician needs to know.

When and how did your illness begin? Did your symptoms begin suddenly (like being hit by a stray bullet)? Or did they come on gradually over a period of weeks, months, or years?

A good history looks at all sorts of things, past and present. And there are dozens and dozens of pertinent things your physician needs to know about you. Here are some of them.

- What do you eat for breakfast, lunch, dinner and snack?
- Do you smoke and drink alcoholic beverages?
- What drugs or other medications do you take?
- What are your present environmental exposures at home or in your workplace?
- Are you bothered by:
  - perfumes and colognes?
  - fabric shop odors?
  - dusts or molds?
  - pollens?
  - animal danders?
  - occupational exposures?
- Your home environment—are you exposed to smoke, insecticides, gas cooking stove, or odorous carpets?
- Do you use lawn chemicals or pesticides outdoors?

To help me evaluate patients who call or write me seeking help, I have for many years said: "Write me a letter. Tell me anything and everything you'd like for me to know. You can't make it too long. Make it a story and include things you may feel aren't relevant."

I've also devised several questionnaires and history forms—including my 70-item yeast questionnaire. As you know, I assigned a high value to certain questions, especially the history of repeated or prolonged courses of broad-spectrum antibiotic drugs for respiratory, urinary, or other infections.

An Empathetic Attitude by the Physician Is Important

Questionnaires help the physician learn more about the patient's story and her concerns. No doubt about it. Yet, they should not serve as a substitute for a careful, one-on-one, private doctor/patient discussion. In my experience, more than anything else, people with yeast-related health problems want to be listened to by a kind, compassionate, empathetic physician.

Physical Examination

A careful physical examination is important. I'm not talking only about sticking a tongue blade into your mouth and asking you to say, "Ah," or listening to your heart and lungs with a stethoscope, feeling your abdomen and carrying out other examinations. I'm also referring to the physician's overall "inspection."

How do you appear to the physician? Depressed? Cheerful? Alert? Drowsy? Agitated? Anxious? Hostile? Apathetic? All of these person-to-person evaluations add to the physician's impressions obtained from your history, detailed physical examination and the laboratory examinations, X-rays, and tests of many types.

Laboratory Examinations and Other Tests and Studies

Sometimes tests can be critically important. Here are a few examples. An X-ray of the chest of a tired patient who has been losing weight may show a tumor. An examination of the urine could show evidence of a hidden infection. Or a SMA screening blood test could show an elevated blood urea nitrogen (BUN)—an evidence of kidney failure. So laboratory studies are important. Yet, in my medical practice over the years, I never used a laboratory alone to "make the diagnosis."

How Is a Diagnosis of a Candida-Related Health Problem Made?

It is based on all three of the diagnostic steps I've just outlined. But of these three steps, history is by far the most important. And when it is combined with the response of the patient to a special diet and antifungal medication, the diagnosis can be confirmed.

Tests for *Candida Albicans*

Even though laboratory studies do not "make the diagnosis" of a candida-related health problem, many physicians I've consulted have found that laboratory studies may help. These include:

- Candida blood studies of various types, including measurement of candida antibodies, antigens and immune complexes (Lab #2, 4, 10, 12, 13, 15 and 17).
- Stool studies for candida and other organisms (Lab #5, 6, 8, 10 and 14).
- Comprehensive examination of the stool (Lab #10 and 14). Such analyses help in the evaluation of patients with candida-related disorders and other chronic complaints.

Tests for Food Allergies and Sensitivities

With few exceptions, people with yeast-related health problems are bothered by sensitivities to foods they're eating every day. Although the commonly used allergy scratch tests often show negative results, other tests may be helpful, including:

The ALCAT test (Lab #1). This patented test measures changes in cell size/volume following incubation of a whole blood specimen with food extracts, food additives, chemicals or molds (including candida) using specialized electronic laboratory instrumentation.

Scientific studies published in the *Annals of Allergy* showed a high correlation of this test with double-blind challenge in food sensitivity. Then a March 1996 presentation at the American Academy of Asthma, Allergy and Immunology showed that the ALCAT test "seems to provide a high efficiency in detecting food additive intolerance."

IgG food sensitivity tests (Lab #11, 14 and 15). These tests for delayed-onset food allergies are designed to detect food-specific immunoglobulin class G (IgC) antibodies. In a recent scientific study using the Immuno Laboratories test, Dr. Sidney Baker concluded:

> Despite negative mitigating factors, the study showed with overwhelming statistical significance, that the avoidance of IgG foods resulted in a decrease in symptom severity greater than that achieved by a placebo diet.[33]

Tests to assess intestinal permeability (Lab #10 and 15). During the past several years, a number of reports in the medical literature have described what has been termed "a leaky gut." The most clinically applicable way of evaluating mucosal integrity of the gastrointestinal tract is the lactulose and mannitol challenge test. This test involves administering a solution of lactulose and mannitol, combined with glycerol, to an individual who has fasted overnight, and then collecting the urine for six hours.

In individuals with altered gastrointestinal mucosa there's an increased absorption of lactulose, which is subsequently secreted in the urine.[34]

My Comments

A number of my physician consultants have found that laboratory tests help in studying patients with chronic health problems, including those that are yeast related. However, until February, 2000, when people asked, "Is this a laboratory test which will help my physician decide if my health problems are yeast-related?," I would answer, "No."

Now, based on information from Drs. R. Scott Heath and Stuart Lanson, candida immune complex tests performed by the AAL Reference Laboratory provide relevant information. According to Heath, 80% of his patients with neurological problems who showed elevated candida immune complexes improved on oral doses of antifungal medications and the level of immune complexes decreased significantly. Twenty percent of patients who did not respond to three months of oral antifungal medication continued to have elevated immune complex levels.

Yet, I hasten to add that individuals with complex medical problems may benefit from other tests, which often provide essential information. Such tests are especially indicated in people with yeast-related disorders who continue to be sick. These include:

- Tests to identify hidden nutrient deficiencies (Lab #8, 15 and 16).
- Tests to assess the cells' ability to prevent free-radical damage (Lab #15 and 16).
- Tests to study the levels of environmental pollutants in the blood and tissues (Lab #9).

Laboratories

1. American Medical Testing Laboratories, 1 Oakwood Blvd., Suite 130, Hollywood, FL 33020 (800-881-2685 or FAX 954-923-2990).
2. AAL Reference Laboratory, 1715 E. Wilshire #715, Santa Ana, CA 92705 (800-522-2611 or FAX 714-972-9979).
3. Metagenics, 4403 Vineland Rd., Suite B12, Orlando, FL 32811 (800-647-6100 or 407-423-0089).
4. Cerodex Lab, Rt. 1 Box 130, Washington, OK 73093 (405-288-2383 or FAX 405-288-2228).
5. Consulting Clinical and Microbiology Laboratories, 333 S.W. 5th Ave., Suite 620-7, Portland, OR 97204 (503-222-5279).
6. Diagnos-Techs, Inc., 6620 S. 192nd Place, Bldg. J-104, Kent, WA 98032 (425-251-0596 or FAX 425-251-0637).
7. Doctors Data, Inc., Box 111, 170 W. Roosevelt Rd., W. Chicago, IL 60185 (800-323-2784).
8. Pantox Laboratories, 4622 Santa Fe St., San Diego, CA 92109 (800-726-8696 or FAX 619-272-1621).
9. Accu-Chem Laboratories, 990 N. Bowser Rd., #800, Richardson, TX 75081 (800-451-0116 or FAX 214-234-5707).
10. Great Smokies Diagnostic Laboratory, 63 Zillicoa Street, Asheville, NC 28801 (800-522-4762 or FAX 704-253-0621).
11. Immuno Laboratories, 1620 W. Oakland Park Blvd., Ft. Lauderdale, FL 33311 (800-231-9197 or FAX 305-739-6563).
12. Immunodiagnostic Laboratory, Inc., 10930 Bigge St., San Leandro, CA 94577 (800-888-1113 or FAX 510-635-5667).
13. Immunosciences Lab, Inc., 8730 Wilshire Blvd., #305, Beverly Hills, CA 90211 (310-657-1077 or FAX 310-657-1053).

14. Meridian Valley Clinical Laboratory, 515 W. Harrison St., Kent, WA 98032 (253-859-8700 or FAX 253-859-1135).
15. Metametrix Medical Research Laboratory, 5000 Peachtree Ind. Blvd., Norcross, GA 30071 (770-446-5483 or FAX 770-441-2237).
16. SpectraCell Laboratories, Inc., 515 Post Oak Blvd., Suite 830, Houston, TX 77027 (800-227-5227 or FAX 713-621-3234).

Yeast-Related Weight Problems

An estimated 30 percent of all Americans weigh more than they would like to weigh. Many of the reasons are known, including insufficient exercise and too much food—especially fats, sugars and simple carbohydrates. Heredity also seems to play a role in many families.

Yet, other causes of being overweight are unknown or poorly understood, in spite of intensive studies by scientists with impeccable academic credentials.

Can weight problems be yeast-related? In my opinion, the answer is "yes" in some people. Here's the story of Adelaide (not her name):

> During childhood and early adolescence I ate what I thought was a reasonably good diet. I went to gym classes, played softball and soccer, but I was never an exercise buff. I never worried about gaining too much weight. In fact, my mother on many occasions told me to eat more because she felt I was too skinny.
>
> I didn't start my periods until I was almost 14 and about the same time I began to be troubled with pimples. When face washing and other local measures didn't seem to clear them up, my mother took me to a dermatologist who put me on tetracycline, which I took on a daily basis for more than two years. By the time I was a senior in high school, I developed other problems, including vaginal yeast infections, PMS, fatigue, constipation, bloating and headache. I also craved sugar. That's about the time I began putting on the weight.
>
> After I stopped taking the tetracycline, my vaginal problems became less frequent, yet I continued to be bothered by other

symptoms, including fatigue, headache, bloating and constipation. And my weight continued to increase.

Moving along with my story, I married at 21, promptly had a baby and then another one. The weight I gained with each pregnancy never seemed to go away. At the age of 28 I weighed 180 lbs., and I was only 5′5″.

Most of my other symptoms continued to bother me, including some new ones. Yet, a series of examinations and tests by several different doctors always came up with negative results. At least the doctors couldn't find anything wrong with me. They said it was stress and that I simply was eating too much.

Then, a friend gave me a copy of *The Yeast Connection*. And as Paul Harvey would say, "Here's the rest of the story."

I found a kind, caring and knowledgeable doctor who put me on nystatin. A nurse nutritionist who worked with him helped me with my diet. She emphasized the importance of staying off sugar, which she said feeds the yeast.

My improvement didn't occur overnight and I had some ups and downs, but today, 11 months after I started my yeast journey, my symptoms are 90 percent better (most of the time). I now weigh 130 and am holding. Thanks to Dr. Truss for his discovery and to you for writing about the "yeast connection."

As you can imagine, I was delighted to get Adelaide's letter. I've received dozens of similar ones from people during the last decade, including Barbara Dewey, a Michigan woman whose letter I included in *The Yeast Connection*. Because of her own experiences, Barbara has been teaching diet workshop classes to help more people with yeast-related problems.

How and why are weight problems sometimes yeast-connected? I don't claim to know all the answers, but here are possible mechanisms:

As pointed out by Dr. C. Orian Truss, individuals with candidiasis develop significant metabolic and biochemical abnormalities. Moreover, Truss and many other observers have noted that sugar craving, fatigue and other manifestations of hormone dysfunction occur almost uniformly in individuals with yeast-related problems.

References

1. Kamen, B., "FOS: A Healthful Sweetener (No kidding!)," *Let's Live,* October 1992; pp. 32–34.

2. Spiegel, J. E., et al, "Safety and Benefits of Fructooligosaccharides as Food Ingredients," *Food Technology,* January 1994; pp. 85–89.

3. Jones, M., "FOS—A 'Good for You' Sweetener!" *Mastering Food Allergies,* November 1992; Vol. 7 No. 10, MAST Enterprises, Inc., 2615 N. 4th St., #616, Coeur d'Alene, ID 83814.

4. Blumenthal, M., Personal communication, February 1996.

5. Ellwein, L. B., Cohen, S. M., "The Health Risks of Saccharin Revisited," *Critical Reviews in Toxicology,* 1990; 20(5):311–326.

6. Roberts, H. J., *Sweet'ner Dearest: Bittersweet Vignettes About Aspartame (NutraSweet),* Sunshine Sentinel Press, P.O. Box 8697, West Palm Beach, FL 33407, 1992.

7. Roberts, H. J., *ASPARTAME (NutraSweet), Is It Safe?* Charles Press Publishers, P.O. Box 15715, Philadelphia, PA 19103.

8. Eisenberg, D. M., et al, "Unconventional Medicine in the United States," *N. Engl. J. Med.,* 1993; 328:246–252.

9. Compiled by the Burton Goldberg Group, *Alternative Medicine: The definitive guide,* Future Medicine Publishing, Inc., Puyallup, WA, 1993.

10. Kleijnen, J. and Knipschild, P., "Ginkgo biloba," *The Lancet,* 1992; 340:1136–1138.

11. Mindell, E., *Earl Mindell's Herb Bible,* Fireside/Simon and Schuster, New York, 1992; pp. 83–85.

12. Galland, L. and Bueno, J., "Advances in Laboratory Diagnosis of Intestinal Parasites," *American Clinical Laboratory,* 1989; pp. 18–19.

13. Lee, M. J., Johnson, J. F., Baskin, W. N. and Barrie, S., "Trends in Intestinal Parasitology. Part II: Commonly Reported Parasites and Therapeutics," *Practical Gastroenterology,* 1992; Vol. 16 No. 10.

14. Gittleman, A. L., *Guess What Came to Dinner—Parasites and your health,* Avery, Garden City Park, NY, 1993.

15. Lin, D. J., *Free Radicals and Disease Prevention,* Keats Publishing Company, New Canaan, CT, 1993. (Reprinted with permission.)

16. Martorano, J., *Unmasking PMS—The complete PMS medical treatment plan,* M. Evans and Company, New York, 1993.

17. Hargrove, J. T., Maxson, W. S., Wentz, A. C. and Burnett, L. S., "Menopausal Hormone Replacement Therapy with Continuous Daily Oral Micronized Estradiol and Progesterone," *Am. J. of Obstet. Gynecol.,* 1989; 73:606–612.

18. Lee, J. R., *Natural Progesterone: the Multiple Roles of a Remarkable Hormone,* BLL Publishing Company, P.O. Box 2068, Sebastapol, CA 95473, 1993.

19. Lee, J. R. and Bland, J., *Preventive Medicine Update,* (audiocassette), HealthComm, Inc., 5800 Soundview Dr., Gig Harbor, WA 98335 (1-800-843-9660).

20. Kamen, B., *Hormone Replacement Therapy, Yes or No?* Nutrition Encounter, Box 5487, Novato, CA 94948, 1993; pp. 109–124.

21. Prior, J. C., Bigna, Y., and Alojada, N., "Progesterone and the Prevention of Osteoporosis," *Canadian Journal of OB/GYN* and *Women's Health Care,* 1991; 3:181.

22. Prior, J. C., et al, "Cyclic medroxyprogesterone treatment increases bone density: a controlled trial in active women with menstrual cycle disturbances," *Am. J. Med.,* 1994; 96(6):521–530.

23. Truss, C. O., *The Missing Diagnosis,* P.O. Box 26508, Birmingham, AL 35226; pp. 30–31.

24. Crook, W. G., *The Yeast Connection,* Third Edition, Professional Books, Jackson, Tennessee and Vintage Books, New York, New York, 1986; pp. 221–222.

25. Iwata, K., "A Review of the Literature on Drunken Symptoms Due to Yeast in the Gastrointestinal Tract," in Iwata, K. (ed), *Yeasts and Yeast-Like Microorganisms in Medical Science,* University of Tokyo Press, Toyko, 1976; pp. 260–268.

26. Edwards, W. J., ".10% Solution—A New Defense for DUI Cases: The Auto-Intoxication Syndrome," *The Champion,* 1993; 17:41–42.

27. Noon, L., "Prophylactic inoculation against hay fever," *The Lancet,* 1911; 1:1572.

28. Liebeskind, A., "*Candida albicans* as an allergenic factor," *Annals of Allergy,* 1962; 20:394–396.

29. Kudelko, N. M., "Allergy and chronic monilia vaginitis," *Annals of Allergy,* 1971; 29:266.

30. Truss, C. O., "Restoration of Immunological Competence to *Candida albicans,*" *J. of Orthomol. Psych.,* 1980; Vol. 9 and *The Missing Diagnosis,* pp. 157–159.

31. Miller, J. B., "Relief of Premenstrual Symptoms, Dysmenorrhea and Contraceptive Tablet Intolerance," *J. Med. Assoc., St. of Alabama,* 1974; 44:57–60.

32. McEwen, L. M., "Enzyme Potentiated Hypodesensitization," *Annals of Allergy,* 1975; 35:98–103.

33. Baker, S. M., McDonnell, M. and Truss, C., "Double-Blind Placebo-Diet Controlled Crossover Study of IgG Food ELISA." Presented at the American Academy of Environmental Medicine Advanced Seminar, Virginia Beach, VA, October 1994.

34. Travis, S., and Menzies, I., "Intestinal Permeability: Functional Assessment and Significance," *Clinical Science,* 1992; 82:471–488.

CHAPTER · THIRTEEN

Potpourri

■ ■ ■

Celiac Sprue and the Yeast Connection

Celiac sprue (CS), is a relatively uncommon inherited disorder that affects digestive metabolism. It manifests itself principally as an inability of the small bowel to digest and absorb wheat and other gluten-containing foods.

In his 1992 book, *Can a Gluten-Free Diet Help? How?*, Dr. Lloyd Rosenvold said:

> I believe in some cases (maybe even many cases) the presence of celiac sprue, which is secondary to gluten intolerance offers a fertile . . . culture medium for the propagation of yeast organisms. In any case of intestinal yeast infection the possibility of a concomitant CS should be considered until it can be confidently ruled out . . .
>
> How are CS and yeast infections connected, if they are? I'll briefly state my simple concept. The putrefactive processes in the bowel, incidental to the maldigestion and malabsorption present as a result of the CS, furnishes . . . an ideal culture medium for the propagation of *Candida albicans*. . . .[1]

Halitosis

In a report in a peer-reviewed medical journal, Dr. Martin Zwerling and associates told of their success in helping 70 out of 79 patients with yeast-related problems, including some who were troubled with bad breath. Here's an excerpt from their article:

In a busy ENT practice it is not difficult to find patients with intractable bad breath, most of whom have had their tonsils and teeth removed, along with assorted abdominal organs, in unsuccessful attempts for relief.

Fifteen patients with halitosis were placed on the therapeutic trial (avoidance of sugar and foods rich in yeast and mold, elimination of antibiotics, steroids and birth control pills . . . and nystatin and/or ketoconazole).

Within 24 to 48 hours all 15 patients reported marked improvement in their breath odor and this was confirmed by their families and the examining physician.[2]

Head-Clearing Effect of Diflucan

Recently Ray C. Wunderlich, Jr., M.D., St. Petersburg, Florida, wrote me about some of his experiences in treating his patients with antifungal medications. Here's an excerpt.

> I've had about ten patients who had various chronic illnesses ranging from arthritis to seizures to MS to gastrointestinal disorders. Upon treatment with Diflucan, 200 mg. a day, a decisive head clearing occurred. In other words, foggy thinking, brain-fag [exhaustion] or spaciness remitted.
>
> Several of these patients had MS. One is an entertainer who travels across the world. None had "classic" symptoms/signs to suggest candidiasis. Upon discontinuing treatment, symptoms usually recur. The longest duration of treatment so far is two years. Why would anyone choose such an expensive placebo.?

Headaches and *Candida Albicans*

In an article in the *Journal of Advancement in Medicine,* Gunnar Heuser, M.D., and colleagues presented their findings on studying 17 patients with recurrent classic or common migraine. All of those selected has a classic or common migraine of more than two years duration, with at least two attacks of headache per month. All had shown resistance to well-established treatment modalities for migraine headache.

These patients were put on a restricted diet and given acidophilus supplements. Four weeks after initiation of the diet they were placed on oral nystatin. In summarizing their observations, these investigators concluded that, "Some vascular headaches have candidiasis as a trigger and respond to appropriate treatment."[3]

Digestive Problems and Garlic

Abdominal pain, constipation and intestinal gas plague many people. And if a person with these symptoms gives a history of repeated or prolonged courses of antibiotic drugs, oral nonprescription antiyeast agents may help relieve their symptoms. Here's a report I received in a letter from Calvin Thrash, M.D., 30 Uchee Pines Rd., #31, Seale, Alabama 36875-5703:

> I saw a 45-year-old lady with a severe yeast problem—almost unable to eat anything, lots of gas, bloating, discomfort, poor digestion, etc. She has tried everything, including nystatin, with minimal help. I told her to take the liquid Kyolic, three teaspoons, three times a day, along with diet, and some other suggestions. She called me three months later to say that within two weeks her symptoms were all gone.

Yeast-Connected Urticaria (Hives)

Some years ago, I read a paper by a prominent allergist who said that he would rather see a tiger walk into his office than a patient with chronic urticaria. He felt this way because finding the cause of this disorder is often impossible. Prednisone, a steroid medication that is often given to relieve symptoms, may promote the growth of *Candida albicans* and set up a vicious cycle of other problems.

The yeast connection to urticaria was clearly described in the medical literature by G. Holti more than 30 years ago. In his group of patients he studied, 27 were clinically cured using oral and vaginal nystatin.[4]

In the spring of 1994, a woman came in to see me complaining of chronic hives. To get an expert opinion on treatment, I called Dr. Robert Skinner of the Department of Dermatology at the University of Tennessee. He told me that many of his patients with chronic hives improved when treated with nystatin or other antiyeast medications.

Lyme Disease and the Yeast Connection

During the past few years, a strange and often puzzling disease has received national attention. It is caused by a microorganism, *Borrelia burgdorferi,* and is transmitted to humans through the bite of infected ticks.

Broad-spectrum antibiotics have been found effective in treating patients with Lyme Disease, especially when given during the acute stage. Other patients given antibiotics for weeks and months continued to experience problems. Here's a possible explanation:

While antibiotics are controlling the microorganisms that cause Lyme Disease, they're knocking out friendly germs in the intestinal tract. As a result, yeasts multiply and cause a weakening of the immune system, leading to persistence of the borrelia infection and other problems.

Suggestions for Obtaining Safer Foods

In an article in *NOHA News,* a newsletter published by the Nutrition for Optimal Health Association, editor Marjorie Fisher commented,

> We can greatly reduce our own and our family's exposure to pesticides by buying organic food. These purchases help society at large, as well as ourselves directly. By creating a demand for the products of the organic farmers, their distributors and the stores selling their products, we're helping those committed to no pesticide use. (Including no insecticides, herbicides, nor fungicides.)
>
> We should, of course, be alert to false claims and deceptive actions. For example, the regular grocery store that advertises a small section of "organic" produce, although they have contaminated the entire store by employing exterminators.
>
> Even with a tiny plot of land we can grow a great deal of our own food organically. . . . From our own garden it can be a delight to eat fresh herbs, vegetables and fruits at the peak of their ripeness and wonderful flavor.

If you'd like a copy of this eight-page newsletter, send a long, self-addressed, stamped (2 stamps), envelope and a donation of $2 to NOHA, P.O. Box 380, Winnetka, IL 60093.

Too Much Iron Can Cause Problems

You do not need an iron supplement unless you're losing a lot of blood during your menstrual periods, or you've had a surgical operation accompanied by extensive blood loss.

According to Dr. Elmer Cranton, "Iron is so potentially dangerous that I recommend blood testing before prescribing it for anyone. Too much iron can shorten your life."

To assess iron status, he recommended the serum ferritin test, and said that the levels should be under 100. He also said:

> You'll be a lot healthier if you're between 50 and 70 if you periodically donate blood. A study was made several years ago of people in this age group who donated blood on a regular basis and others who did not. *The death rate of those who donated blood was half of that observed in those who did not.*[5]

Why does iron cause trouble? Although all the answers aren't known, it may have to do with free radicals, and iron appears to increase free radical damage. (See pages 190–193.)

More On Rotation Diets

Along with many physicians interested in hidden food allergies, I've found that my allergic patients who rotate their diets usually get along well and develop fewer new food allergies.

Rotating your diet means eating a food only once every four to seven days. For example, if you're allergic to egg and after avoiding it for several months, you eat an egg and it doesn't bother you, you can try eating an egg once a week and see if you tolerate it. You can do the same with other foods.

Here's a summary of the advantages of rotating your foods:

- It helps you maintain your tolerance to the foods you can eat now, greatly lessening your chances of becoming allergic to other foods.
- It helps in the treatment of current food allergies.
- It aids in identifying foods that could be causing your problems.

You can find more information about rotated diets in my 1989 book, *The Yeast Connection Cookbook,* which was co-authored by Marjorie Hurt Jones, R.N., and in the following publications:

- Sondra K. Lewis' 1996 book, *Allergy and Candida Cooking—Made Easy,* Canary Connect Publications, P.O. Box 5317, Coralville, IA 52241-0317.
- Nicolette M. Dumke's book, *Allergy Cooking With Ease,* Starburst Publishers, P.O. Box 4123, Lancaster, PA 17604, 1992.
- Sally Rockwell's, *Cooking With Candida Cookbook* (revised), P.O. Box 13056, Seattle, WA 98103.

Pesticides and Breast Cancer

A number of scientific studies show that women who are exposed to DDT and other chemicals are much more apt to develop breast cancer than other women. Although DDT has been banned in the United States, it is widely used in other countries and should be a cause of concern to all of us.

In a letter to the editor published in the *Journal of the American Medical Association,* David Perlmutter, a Naples, Florida, neurologist, commented on the pesticides in imported fruits and vegetables. In his letter he pointed out that the General Agreement on Tariffs and Trades (GATT) was designed to stimulate worldwide agricultural trade. Yet, he pointed out that the GATT rules could allow substantial levels of pesticide residues on import produce. And he said,

> Breast cancer, which now affects one in eight American women, must truly be regarded as a modern epidemic. This year 180,000 American women will be diagnosed as having this disease and a third of them will die of it. . . . Recent evidence strongly supports the relationship between tissue levels of organochlorines (like DDT) and the incidence of breast cancer . . .
>
> While recognizing the importance of national economics, this agreement, (GATT) which affects safety standards of imported produce, demonstrates that the health of American women is not a primary concern.[6]

Hormone Copycats

I first heard that dioxin and other chemicals could affect hormones from Mary Lou Ballweg of the Endometriosis Association. A short time later I learned about the work of Theo Colburn, Ph.D., and the National Wildlife Federation. This organization has spearheaded research studies, which have described the frightening effect of dioxin and other chemicals on humans and animals.

Some of the most disturbing evidence has come from wildlife, including alligators, fish, turtles and birds. Yet, the same chemicals which are affecting animals, also affects humans.

To obtain a copy of the 61-page report, "Hormone Copycats," send $6 to National Wildlife Federation, Great Lakes Natural Resource Center, 506 E. Liberty St., Ann Arbor, MI 48104. Fax: 734-769-3351.

Does Your Stomach Burn?

During the past decade, countless people who experience burning in the upper part of their abdomens have taken medications, such as Tagamet and Zantac. Although these medicines lessen acid secretion and relieve symptoms, recent research studies appear to provide better answers.

The research I'm talking about was originated by Barry Marshall, M.D., a young Australian/American physician. Marshall found that the bacterium *Helicobacter pylori* played a major role in causing gastritis and gastric ulcers. He noted that by taking a prescription of Flagyl, tetracycline and Pepto-Bismol, more than 80 percent of ulcer patients are cured.[7]

Dr. Marshall's discovery is exciting. The treatment program he outlined is helping many people. Yet, since tetracycline and Flagyl encourage the growth of *Candida albicans* in the digestive tract, adding nystatin and a probiotic to the treatment program is an option worth considering.

Medical Research: What Should You Believe?

This was the title of a cover story in *People's Medical Society Newsletter.* The article said that often reports in medical journals, which are discussed in the press and media, provide consumers with conflicting advice. In this article, editor Charles B. Inlander and staff pointed out that there can be flaws in the design, differences in the overall health of study participants and differences in the way data is analyzed. In summarizing their thoughts about medical research, they said:

> The bottom line is that medical research is not perfect . . . When it comes to medical research, what should you believe?
>
> Not everything. Not nothing. You should evaluate each study for its individual merits, take it as food for thought. But don't make rash medical decisions or change your lifestyle solely because of a single study.[8]

Indoor Plants Absorb Pollutants

Research I've read about in the past several years show that healthy indoor plants may absorb many indoor air pollutants. One report stated that two healthy growing plants (about two feet high in a 10x10-foot room) will absorb and destroy most volatile pollutants. Plants do this through the leaves, soil and roots. The beneficial microorganisms in the soil do their part in breaking down the gasses.

Eliminating Toxic Chemicals from Your Body

Formaldehyde, trichlorethylene, pesticides and other toxic chemicals may play a part in making you sick. The presence of these toxins can now be determined by laboratory studies. If high levels are found, detoxification can be carried out.

For more information about "detox" programs, write to William Rea, M.D., 8345 Walnut Hill Lane, Suite 205, Dallas, TX

75231 or Allan Lieberman, M.D., Center for Ecological Medicine, 7510 Northforest Dr., North Charleston, SC 29418.

Headaches and the Herb Feverfew

In a report in *USA Weekend*, Jim Duke, Ph.D., a leading expert on herbs, cited a report in the Harvard Medical School Letter that said eating feverfew leaves has become a familiar method for preventing migraine attacks in modern England. You can obtain more information about feverfew and other herbs from the American Botanical Council, P.O. Box 201660, Austin, TX 78720; Fax: 512-331-1924.*

Looking for an Alternate Health Practitioner?

In November 1991, the editors of *Let's Live Magazine* published the results of a Health Care Survey. They found that at least 43 percent of the respondents would consider seeing an alternative medicine* doctor, or are planning to consult one in the near future. To find such a practitioner, write or call one of the following organizations:

- The American Association of Naturopathic Physicians in Seattle, Washington (206-298-0126). Leave your name and complete address and you will receive a list of naturopathic doctors in your area. There is a $5.00 fee for information.
- The National Center for Homeopathy, 7630 Fay Avenue, La-Jolla, CA 92067 (619-551-7788).
- Center for Mind/Body Medicine, 1110 Camino del Mar, Suite G, Del Mar, CA 92014 (619-794-2425).
- American Massage Therapy Association, 820 Davis St., Suite 100, Evanston, IL 60201 (847-864-0123).

References

1. Rosenvold, L., *Can A Gluten-Free Diet Help? How?,* Keats Publishing, New Canaan, CT, 1992; pp. 28–32.

---

*See also pages 261–263.

2. Zwerling, M. H., Owens, K. N., Ruth, N. H., "Think Yeast—The Expanding Spectrum of Candidiasis," *J. South Carolina Med. Assoc.*, 1984; 80:454–456.

3. Heuser, G., Heuser, S., Rahimian, P. and Vojdani, A., *J. of Advan. in Med.*, 1992; 5:177–188.

4. Holti, G., in Symposium on Candida Infection, edited by Winner, H. L. and Hurley, R., Edinburg: Livingstone, 1966; pp. 73–81.

5. Cranton, E., Personal communication, July, 1995.

6. Perlmutter, Letter to the Editor, *JAMA*, Vol. 71, April 20, 1994.

7. SerVaas, C., "Search for the Cure Continues," *The Saturday Evening Post*, September/October 1994 and *Saturday Evening Post Health Update Newsletter*, Summer 1994, The Saturday Evening Post Society, 1100 Waterway Blvd., Indianapolis, IN 46202.

8. *The People's Medical Society Newsletter*, October 1994, Vol. 13 No. 5, 462 Walnut St., Allentown, PA 18102.

CHAPTER · FOURTEEN

Sugar-Free, Yeast-Free Recipes*

Swiss Steak

2 lbs. round steak, cut into serving-size pieces
2 tbsp. unrefined vegetable oil
1 medium onion (1/2 cup), chopped
1/2 cup chopped celery
1/4 cup chopped green pepper
1 cup peeled, chopped fresh tomatoes
1/2 cup carrots
2 cups water

Heat skillet to 370°, add oil, then brown the meat. Place in casserole dish and add remaining ingredients. Cover and bake at 300° for 3 to 4 hours. Serves 4.

Meat Loaf

2 lbs. ground beef
1 cup oats, uncooked (old-fashioned)
1 egg, beaten
1/4 cup chopped onion
1/2 cup water
1 tsp. salt

Combine ingredients. Shape into loaf and bake in 9x5 pan at 350° for about 1 hour. Serves 6.

*You'll find over 200 recipes in *The Yeast Connection Cookbook*.

Sauteed Liver Slivers

1/4 cup whole wheat flour
1/4 cup cold-pressed vegetable oil
1 lb. beef liver, cut into slivers
1 onion, chopped
Salt to taste

Roll liver slivers in whole wheat flour. Heat oil in pan. Saute onions and liver. Serves 4.

Fish Cakes

2 cups tuna fish
1 cup cooked brown rice
1 egg
1 small onion, finely chopped
1 tbsp. lemon juice (fresh)
1 tbsp. melted butter or unrefined vegetable oil
1 cup finely ground nuts

Beat egg; add tuna, rice, onion, lemon juice and oil. Blend well and form into patties. Roll in nut meal and place on oiled baking sheet. Bake in 350° oven about 15 minutes or until brown and bubbly. May substitute any cooked fish or poultry for tuna. Makes approximately 8 patties. Serves 4.

Pork Chops and Brown Rice

8 pork chops
1 cup brown rice
2 1/2 cups boiling water
1/2 cup chopped green pepper
1/4 cup chopped onion
1/2 tsp. salt

Soak rice in boiling water for at least 30 minutes. Brown pork chops. Set aside. In same frying pan add remaining ingredients. Simmer 10 minutes. Pour rice mixture into an oiled baking dish. Layer chops on top. Cover. Bake at 350° for 1 hour. Serves 4.

Baked Chicken

Chicken, cut into serving-size pieces
Cold-pressed vegetable oil
Salt

Oil bottom of oven-baking dish. Lay chicken pieces in pan (skin-side down), sprinkle with salt and cover with foil. Bake at 375° for 1/2 hour. Remove foil, turn pieces, and bake an additional 20 to 30 minutes, until brown and tender.

Pork Chop Casserole

1 10-oz. package frozen green beans
2 medium onions, sliced
2 or 3 potatoes, peeled and sliced
6 pork chops
Salt

Layer beans, onions, potatoes in an oiled 13x9 baking pan or casserole. Sprinkle with a bit of salt. Place pork chops on top, cover and bake at 375° for 3/4 hour. Remove cover (or foil) and bake an additional 15 minutes until chops are tender and brown. Serves 4.

Chicken Salad

2 cups finely chopped cooked chicken
1/2 cup finely chopped celery
2 to 4 hard cooked, chopped eggs
1 medium onion, chopped

Moisten with sugar/honey free mayonnaise obtainable from your health food store. Serves 4.

Easy Chicken and Rice

3 lbs. frying chicken pieces
1 cup brown rice
2 cups water
1/2 tsp. salt
1 1/2 tbsp. butter
3 tbsp. chopped fresh parsley
Optional: Onions, celery, green pepper, nuts

Place rice, water, salt, butter and parsley in 4-quart casserole. Stir and bring to a boil. Salt chicken and lay on top of rice. Lower heat to simmer; cover tightly and cook 45 to 60 minutes until water is absorbed and chicken is tender. Serves 4.

Barley or Rice Soup

1–3 lbs. chicken, disjointed
8 cups cold water
1/2 cup onion, chopped
1 tbsp. sea salt
1/2 cup barley or brown rice
1 cup celery, chopped

Place all ingredients in a large pan and cook over medium heat until done. Or, may be cooked all day in a slow cooker. Serves 4.

Best Barley Soup

1/4 cup hulled barley
1 cup carrots
1/2 cup celery
1/4 cup onions, chopped
2 cups tomatoes, chopped
1 cup peas—fresh or frozen
Fresh parsley

Cook barley in 6 cups of water for 1 hour. Add remaining ingredients and cook until tender. Add parsley just before serving. Serves 4.

Salmon Patties

1 lb. canned salmon with liquid
1/3 cup whole wheat flour
1/4 cup stoneground corn meal
1/4 cup wheat germ
2 eggs, beaten
1/2 cup chopped onion
1/4 cup chopped bell pepper
1 tbsp. lemon juice
1/4 to 1/2 cup unrefined vegetable oil for frying

Flake salmon, mashing bones well. Mix all ingredients (except for the oil) together. Form into patties. Brown in oil on medium heat for 15 minutes. Serves 4.

Stewed Okra

Okra
Onion
Tomatoes
Corn
Sea salt

Saute okra and onion in a bit of unrefined vegetable oil. Add fresh tomato, fresh or frozen corn, a bit of salt and simmer until tender. May thicken with stoneground corn meal if desired.

Zucchini and Tomatoes

Zucchini
Onion
Tomato
Fresh Parsley
Sea salt

Saute zucchini and onion until tender. Add chopped tomato and continue to stir until cooked. Add parsley and a bit of sea salt and serve.

Green Beans with Almonds

Fresh or frozen green beans
Onion
Unrefined vegetable oil
Fresh parsley
Slivered blanched almonds
Salt

Steam green beans until tender. In separate pan saute onion in vegetable oil until tender. Add green beans, parsley and almonds. Sprinkle lightly with salt and serve.

Summertime Salad

1 cup cucumber, diced
1/4 cup onion, chopped
1 cup tomatoes, diced
1/2 tsp. sea salt (or less)

NO NEED FOR DRESSING. Serves 4.

Acorn Squash

Squash
1 tbsp. unrefined vegetable oil
1 cup water
Sea salt

Cut squash in half and remove seeds. Place oil and water in baking pan. Lay halves in pan, cut-side down, and bake in 370° oven for 30 minutes. Turn (face up), brush with oil and sea salt and bake another 20 to 30 minutes until tender.

Stir-Fried Vegetable Scramble

2 tbsp. unrefined vegetable oil or butter
2 tbsp. chopped onion
2 tbsp. chopped green pepper

1/2 cup fresh chopped tomato
1/2 to 1 cup cooked vegetables
2 to 4 slightly beaten eggs

Heat skillet, add oil, onions and green peppers. Stir-fry until tender. Add tomato and other vegetables. Bring to boil, stirring constantly. Add eggs and cook, stirring gently. Serve immediately.

Wheat Biscuits

2 cups whole wheat flour
4 tsps. baking powder
1/2 tsp. salt
1/3 cup unrefined vegetable oil
3/4 to 1 cup water or milk

Mix together dry ingredients, add oil and mix well (important). Add enough water to make a soft dough that is not sticky. Mix just enough to moisten dry ingredients. With hands, pat out dough to 3/4-inch thickness on floured board. Cut with glass, place on oiled baking sheet and bake 20 minutes (or until done) in a 450° oven. Makes 12 biscuits.

Zucchini Italiano

6 small zucchini, cut into 1/4-inch slices
2 tbsp. butter
1 cup water or milk
3 slightly beaten eggs
1 tsp. salt
2 tsp. basil

Put zucchini in 1 1/2-qt. casserole. Dot with butter. Bake at 400° for 15 minutes. Combine and pour remaining ingredients over zucchini. Set casserole in shallow pan with one inch hot water. Bake 350° for 40 minutes or until knife comes out clean.

Whole Wheat Popovers

3 eggs
1 cup water or milk
3 tbsp. butter
1 cup whole wheat flour

Beat eggs until foamy. Add water and butter and continue beating. Add flour and blend. Fill greased muffin cups 2/3 full. Bake at 375° for 50 minutes. Makes 12 popovers.

Banana-Oat Cake

2 cups oat flour
1/4 tsp. salt
1/2 cup mashed banana
2 tsp. baking powder
2 eggs
3 tbsp. cold water
2 tbsp. unrefined vegetable oil

Mix dry ingredients. Beat eggs. Add water, oil and mashed banana. Blend with dry ingredients. Put the batter in a greased 8x8 pan and bake at 350° for 25 to 30 minutes in a preheated oven. Cut in 9 pieces.

Better Butter

1 stick (1/2 cup) butter, room temperature
1/4 cup flaxseed oil

Blend until light and fluffy. Store in glass container. Use as you would regular butter or margarine.

Corn Bread

1 3/4 cups stoneground corn meal
1/3 cup whole wheat flour

3 tsp. baking powder
1/2 tsp. sea salt
1 egg, lightly beaten
3 tbsp. cold-pressed vegetable oil
1 1/2 cups water or milk

Combine dry ingredients, beat egg, add oil and water and blend all together. Bake in oiled 8-inch square pan at 425°, 20 to 25 minutes. Cut in 9 pieces.

Spoon Bread

1 1/2 cups water or milk
3/4 cup stoneground cornmeal
1/2 tsp. sea salt
3 eggs, separated
3 tbsp. cold-pressed vegetable oil

Bring water to boil. Pour in cornmeal and stir constantly and cook until thickened. Add salt and lightly beaten egg yolks and oil. Remove from heat and cool slightly. Fold in stiffly-beaten egg whites. Bake in greased 8-inch square baking dish at 375°, 35 to 40 minutes. Cut in 9 squares.

Rice-Oat Pancakes

1/2 cup rice flour
3/4 cup old-fashioned oats
1 1/2 tsp. baking powder
1/4 tsp. baking soda
2 tbsp. unrefined vegetable oil
1 1/2 tsp. sea salt
3 tbsp. apricot puree (or egg)
1/2 cup soy milk (sugar free)

Blend oats in blender until fine. Mix all ingredients together and drop tablespoonsful into oiled skillet.

Sunflower Crackers

1 cup whole wheat flour
3 tbsp. sunflower seed butter
2 tbsp. cold-pressed vegetable oil
3 tbsp. water
1/4 tsp. sea salt

Blend sunflower seeds in food processor. Combine flour, sunflower seed butter and oil. Gradually add water, adding just enough to form a soft dough. Add salt. Knead and roll on floured surface to 1/8-inch thickness. Cut in shapes, prick with a fork and bake at 350° for 10 minutes or until browned. Cool.

Oat Griddle Cakes

3/4 cup cooked old-fashioned oats
1 1/4 to 1 1/2 cup milk or water
1 egg
2 tbsp. pressed oil
3/4 cup oat flour
1 tsp. baking powder
1/2 tsp. salt

Mix all ingredients together and spoon into a preheated oiled skillet.

Potato Pancakes

3 cups potatoes, grated raw
1 cup onion, grated
3 eggs, beaten
1/2 tsp. sea salt
2 tbsp. whole wheat flour
2 tbsp. unrefined vegetable oil

Drop by tablespoonsful in hot-oiled skillet. Lower heat slightly and brown, turn and brown other side. Makes 12 pancakes.

Rice Muffins

1 1/2 cups rice flour
1/2 tsp. sea salt
2 tsp. baking powder
1/4 tsp. baking soda
4 tbsp. unrefined vegetable oil
3 tbsp. apricot puree
1 cup water or soy milk (sugar-free)

Mix and spoon into greased muffin tins and bake at 350° for 15 to 20 minutes. Makes 12 muffins.

Pancakes

1 1/2 to 2 cups water or milk
2 eggs
2 tbsp. unrefined vegetable oil
1/2 tsp. salt
2 cups whole wheat flour
2 tsp. baking powder

Beat eggs, add water and oil and blend. Add dry ingredients and blend well. Bake on hot, oiled griddle. Makes about 2 dozen 4-inch cakes.

Corn Muffins

1 1/4 cup whole wheat flour
3/4 cup stoneground corn meal
4 1/2 tsp. baking powder
1 tsp. sea salt (optional)
1 egg
2/3 cup water or milk
1/3 cup unrefined vegetable oil

Sift dry ingredients together. In separate bowl blend egg, water and oil; add to dry mixture and stir with a spoon just enough to dampen all ingredients. Spoon into oiled cups of muffin tin. Bake in 425° oven 18 to 20 minutes. Yields 12 muffins.

APPENDIX · A

A Special Message to the Physician

■ ■ ■

If you're like most physicians, you've probably heard about "the yeast connection." And, if you read the *Position Statement* of the American Academy of Allergy and Immunology, you may feel that the relationship of superficial *Candida albicans* infections to a number of chronic health disorders is "speculative and unproven."

You may ask (as have many physicians), "Why haven't scientific studies been carried out to prove the yeast/human interaction, which you so enthusiastically describe in your books? . . . Why haven't you published your observations in peer-reviewed journals?"

You may also say, "I read the report by Dismukes and associates published in the *New England Journal of Medicine.*[1] These investigators described women with recurrent vaginitis, fatigue, headache and other symptoms. And they stated that the women in a control group responded just as well as those treated with oral and vaginal nystatin. So it seems to me that the relationship of yeast infections to chronic illness is unproven."

I can understand your point of view and your criticisms. They obviously have merit. Yet, there are always two (or more) sides to any controversial issue. Here are several of them.

Flaws in the Dismukes Study

Comments by John E. Bennett, M.D.

In an editorial that accompanied the Dismukes article, Bennett, a mycologist at the National Institute of Allergy and Infectious Diseases, said:

Few illnesses have sparked as much hostility between the medical community and a segment of the lay public as the chronic candidiasis syndrome . . . Those who argue for the existence of (this) syndrome . . . have leveled a serious charge against the medical community, claiming it is not fulfilling one of its most important obligations to its patients. This charge is simply put: *You physicians are not listening to your patients* (emphasis added).

Those who argue for the existence of the chronic candidiasis syndrome will complain that diet was not controlled and that it is an important aspect of treatment . . . In fact, none of the proponents of the syndrome have recommended the use of nystatin alone, and are not likely to consider the Dismukes' study an adequate test of their hypothesis.

In the concluding sentence of his editorial, Bennett said:

Additional scientifically sound studies will be needed to determine whether this syndrome does or does not exist; and if it does, what the optimal treatment is for patients.[2]

I'm happy to report that several studies now provide some of the scientific support that the skeptics have been demanding, and other studies are under way.

Comments About Yeast-Related Disorders by Other Physicians

In a statement discussing candida-related illness, Douglas H. Sandberg, M.D., Professor of Pediatrics, Division of Gastroenterology and Nutrition at the University of Miami, said:

Confirmation of the diagnosis remains difficult, evaluation of efficacy of therapeutic measures incomplete; and tools for monitoring a therapeutic response are below the standards we've come to expect in modern medical practice.

In spite of these shortcomings, *I'm convinced that this disorder exists and that it is important. It must be considered in differential diagnosis of patients with a variety of chronic complaints. Since diagnosis at times can be made only through determining response to a therapeutic trial, some patients would have to be treated without a firm diagnosis prior to institution of therapy.*[3] (emphasis added)

Recently, James H. Brodsky, M.D., a diplomate of the American Board of Internal Medicine, a member of the American College of Physicians, and a member of the clinical faculty of Georgetown University Medical School, commented:

> Since my introduction to the relationship between yeast and human illness in the early 1980s, I've seen well over 1,000 patients with some form of yeast-related illness . . . I maintain a general internal medicine practice and make hospital rounds daily. While I find all aspects of my practice fulfilling, nothing has been so rewarding as helping patients with yeast-related illnesses who have been unable to find help elsewhere.[4]

Clinical Reports of the Effectiveness of a Therapy Often Precede Scientific Studies

Clinical reports that describe the effectiveness of a particular method of therapy may precede by decades (or even centuries) the scientific studies, which provide support for the therapy.

More than 10 years ago, two physicians from the University of New Mexico School of Medicine, in an article published in the *Journal of the American Medical Association,* pointed out that an effective treatment for a particular disease is often ignored or rejected because the reasons why the therapy worked aren't understood. *And they said that the only three issues that matter in picking a therapy are: Does it help? How toxic is it? How much does it cost?*[5]

Several thousand physicians in practice and a handful of academicians have found that a sugar-free special diet and nystatin, ketoconazole (Nizoral), fluconazole (Diflucan) or itraconazole (Sporanox) are effective in treating patients with a diverse group of health problems. These range from PMS, chronic fatigue syndrome, interstitial cystitis and psoriasis in adults to recurrent ear infections (and other respiratory infections) and the subsequent development of hyperactivity, attention deficits and autism in children.

I hope you'll take a careful look at the relationship of superficial yeast infections to chronic health disorders, which affect people of all ages and both sexes. Included especially are premenopausal women who feel "sick all over" and other family members who give a history of repeated courses of broad spectrum antibiotic drugs. I feel that in so doing you'll be able to help many of your difficult patients and at the same time make your own practice more interesting and rewarding.

References

1. Dismukes, W. E., Way, J. S., Lee, J. Y., Dockery, B. K., Hain, J. D., "A randomized double-blind trial of nystatin therapy for the candidiasis hypersensitivity syndrome." *N. Engl. J. Med.,* 1990; 323:1717–1723.

2. Bennett, J. E., "Searching for the Yeast Connection," *N. Engl. J. Med.,* 1990; 323:1766–1767.

3. Sandberg, D. H., Statement, "Candida-Related Illness," September 22, 1989.

4. Brodsky, J. H., Statement, "The Importance of Candida-Related Health Problems," October 14, 1993.

5. Goodwin, J. S. and Goodwin, J. M., "The Tomato Effect," *JAMA,* 1984, 251:2287–2290.

APPENDIX · B

Other Sources of Information

· · ·

Health professionals who *may* be interested in treating people with yeast-related health problems include medical doctors (M.D.s and D.O.s), nurse practitioners (N.P.s), advanced practice nurses (A.P.N.s), naturopathic physicians (N.D.s), and chiropractic physicians (D.C.s).

Medical Organizations

American Academy of Environmental Medicine (AAEM)

Physicians in this organization (including M.D.s and D.O.s) are concerned with adverse reactions experienced by individuals who have been exposed to environmental excitants. The resulting disorder, as determined by the person's susceptibility, is termed environmental illness. Excitants to which individual susceptibility exists are found in air, water, food, drugs, and in the home, work and play environments.

Members of this organization are also interested in helping patients with food sensitivities and nutritional deficits. For names of physicians in your area, send a $3 donation and a self-addressed, stamped envelope to AAEM, 7701 E. Kellogg, Suite 625, Wichita, KS 67207.

American College for Advancement in Medicine (ACAM)

Physicians in this society are dedicated to educating physicians on the latest findings and emerging procedures in preventive/nutritional medicine. ACAM's goals are to improve physician skills, knowledge and diagnostic procedures and to develop public awareness of alternative methods of medical treatment.

For names of physicians in your area, send a self-addressed, stamped envelope (2 stamps), to ACAM, 23121 Verdugo Drive #204, P.O. Box 3427, Laguna Hills, CA 92654.

American Holistic Medical Association (AHMA)

Members of this organization (including M.D.s, D.O.s, students studying for those degrees and other licensed health practitioners) emphasize the importance of the whole person—body, mind and spirit—and the interdependence of each of these parts. A cooperative relationship between practitioner and patient is stressed with encouragement that both parties participate fully in health care decisions.

A national physician referral directory lists the names of all AHMA member physicians who are currently accepting new patients. Each listing contains a statement from the physician explaining his/her specialty.

For names of AHMA health professionals in your area, and three association brochures, please send a donation of $10.00 to AHMA, 672 Old McLean Village Drive, McLean, VA 22101.

Other Licensed Health Professionals

During medical school, internship and residency training, most M.D.s and D.O.s have been taught to recognize and treat disease. Little emphasis was placed on promoting health. Although curriculum changes are being made in many medical schools, most M.D.s and D.O.s in practice today received little training in health promotion. In many states other health professionals are now filling this gap. Here are several of them.

Advanced Practice Nurses (A.P.N.s)

During the past two decades, registered nurses with advanced education and experience have become important members of the health care team. Included among these professional nurses are many specialists and subspecialists. Nurse practitioners (N.P.s) are one of these specialists.

According to an article in the January 1995 issue of *The Nurse Practitioner—The American Journal of Primary Health Care,* N.P.s and A.P.N.s are licensed to provide primary health care to patients in all 50 states. Here are excerpts.

1. There are large crossover areas of expertise and overlapping competencies between A.P.N.s and M.D.s;
2. There are many medical and health problems that do not require the experience or expertise of an M.D. to safely provide; and
3. When M.D.s and N.P.s work together collaboratively, the patients receive the best standard of care because each profession has its own unique qualities—the N.P. listens better, manages the care in a broader and more efficient way and is more available while the M.D. has more specialized knowledge for the more serious and complicated care situations.

This article pointed out that care by A.P.N.s is "cost effective, 'patient friendly,' and as safe as care delivered by physicians."

Recently Linda J. Pearson, RN, MSN, FNP, Editor-in-Chief of *The Nurse Practitioner* sent me updated information which was published in the January 1998 issue of *The Nurse Practitioner.*

In 18 states, NPs can prescribe independent of any required physician involvement in the actual prescription writing, and in an additional 31 states NPs can prescribe with some degree of physician involvement or delegation of prescription writing.

For the names of NPs and APNs in your area, contact the American College of Nurse Practitioners (formerly the National Coalition of Nurse Practitioners), 503 Capitol Court Northeast, #300, Washington, DC 20002. Tel. 202-546-4825, Fax: 202-546-4797.

Naturopathic Physicians (N.D.s)

Naturopathy is a method of healing that employs various natural means to empower an individual to achieve the highest level of health possible. Besides providing recommendations on lifestyle, diet and exercise, N.D.s may elect to utilize a variety of natural healing techniques.

These include clinical nutrition, herbal medicine, homeopathy, oriental medicine, acupuncture, hydrotherapy, physical medicine (including massage and therapeutic manipulation, counseling and other psychotherapies), and minor surgery.

Naturopaths practice in most states and Canadian provinces. In several states they are licensed to write prescriptions for naturally derived drugs, including antibiotics and nystatin.

For more information and a referral service, contact The American Association of Naturopathic Physicians, 601 Valley Drive #100, Seattle, WA 98109; phone number 206-298-0126. There is a $5.00 fee for referral information.

Chiropractic Physicians (D.C.s)

More than 50,000 D.C.s practice in the United States and Canada. Some of these licensed health professionals restrict their practices to treatment of neuromusculoskeletal and orthopedic conditions. However, the practice of a substantial percentage of chiropractors may include clinical nutrition, dietary therapy, exercise training, lifestyle modification, environmental control, and mind/body techniques.

They are also trained in diagnosis and laboratory assessment, and commonly work with other specialists when necessary. Some also have experience in helping patients with yeast-related illnesses.

International Health Foundation

After the publication of *The Yeast Connection* in 1983, I received tens of thousands of letters and phone calls from people seeking a physician to help them. To respond to this need, the International Health Foundation (IHF) was incorporated in Tennessee in 1985. The Internal Revenue Service granted the IHF nonprofit, tax exempt status in 1986.

During the past decade, this foundation received more than 70,000 letters from people seeking a physician and other information. In response, IHF established a roster of physicians and sent people a list of physicians in their area who were interested in yeast-related disorders.

Although some people who wrote IHF were able to find a physician through the IHF lists, this approach has left much to be desired. Here are a few of the reasons. In spite of our persistent efforts, fewer than 800 physicians in the United States are listed on the IHF referral roster. Here are other limitations:

- Many physicians listed are so busy that the waiting period to obtain an appointment may be many months.
- The offices of physicians listed may be hundreds of miles from you.
- Physicians listed may not be a member of your health insurance plan.

Although you *may* be able to find help from a physician on the IHF list, *your personal physician may be your best option.* Here's a suggestion which many people who have written and called me have found effective.

If your personal physician is kind and caring, although skeptical of the relationship of yeast to your health problems, write her/him a letter and say,

> Thank you for the patience you've shown in listening to my complaints and for the help you've given me. Yet, in spite of a number of tests and therapies which I've received, I'm continuing to experience health problems that I feel may be yeast-related. Will you work with me?
>
> I realize that you may not believe that yeasts play a part in causing my symptoms. I can understand your point of view. *There is, however, new scientific support of the relationship of yeast to a number of chronic illnesses.*
>
> There are also reports of the effectiveness of nystatin, Diflucan, Lamisil, Sporanox and Nizoral in helping people with a diverse group of chronic illnesses.

In our efforts to respond to the people who write and call IHF seeking help for their own health problems or for their children's health problems, in the summer and fall of 1999 IHF completed the following new publications:

1. A 36-page booklet for adults which includes information about diets, medications and a list of 800 health professionals who have expressed an interest in treating patients with yeast-related disorders.
2. A 76-page booklet, *Children's Health Problems.* In this booklet I discuss:
   - the child with recurring ear disorders
   - the child with "hidden" food allergies
   - the child who keeps a cold
   - the child who is tired and complains of headaches, stomachaches and leg aches
   - the child with autism and other serious developmental problems

   Although I don't claim to possess a "quick fix" for any or all of these problems, information I acquired from many sources shows that they are often yeast-connected.
3. A 24-page booklet, *A Special Message to the Health Professional.* Included in this booklet is a discussion of the yeast controversy and reports from the mid- and late-1990s which support the relationship of yeast infections to many chronic health disorders. These include: asthma, autism, chronic fatigue syndrome, endometriosis, fibromyalgia, headaches, interstitial cystitis, multiple sclerosis and psoriasis.

To obtain booklets 1 and 3 or 2 and 3, send a tax deductible donation of $25 to IHF, Box 3494, Jackson, TN 38303–3494. *If you aren't pleased with the information in these booklets, your money will be refunded.*

You can also find additional information about me and the International Health Foundation on the Internet at www.candida-yeast.com. If you have questions, you may call me on the IHF hotline (901-660-7090) most Tuesdays between 1:15 and 2:15 P.M. CST.

Other Organizations and Support Groups

A number of organizations and support groups in the United States, Canada, England and other countries provide information and help for people with chronic health disorders. These include multiple chemical sensitivity syndrome (MCSS), chronic fatigue syndrome (CFS/CFIDS), fibromyalgia syndrome (FMS) and many others.

By writing to these organizations you may be able to obtain information that may help you. Many are nonprofit organizations and are staffed in part by volunteers. When writing enclose a stamped, self-addressed envelope. A small donation to help cover costs also would be appreciated.

Autism Research Institute
4182 Adams St.
San Diego, CA 92116
Bernard Rimland, Ph.D.
Fax: 619-563-6840

The Burston Clinic
For information on
Candidiasis and all
chronic illnesses
77 Bloor St. West
Main Floor
Toronto, Ontario
M5S 1M2
Phone: 416-944-3526

Candida Allergy
Support Group
Elena McHerron
15 Wildwood Drive
Poughkeepsie, NY 12603
Phone: 914-462-2449

Candida and Dysbiosis
Information Foundation
Elizabeth Naugle, Director
P.O. Box Drawer JF
College Station, TX 77841
Phone: 409-694-8687

Fibromyalgia Network
Kristin Thorson, Editor
P.O. Box 31750
Tucson, AZ 85751-1750
Phone: 800-853-2929
Fax: 602-290-5550

Human Ecology Action
League (HEAL)
P.O. Box 29629
Atlanta, GA 30359
Phone: 404-248-1898
Fax: 404-248-0162

Hyperactivity Helpline
P.O. Box 10085
Arlington, VA 30359
Phone: 703-524-5566

National Vulvodynia
Association
P.O. Box 4491
Silver Spring, MD 29018
Phone: 301-299-0775
Fax: 301-299-3999

New York HEAL
506 E. 84th St.
New York, NY 10028
Phone: 212-517-5937

Price-Pottenger Nutrition Foundation
P.O. Box 2614
La Mesa, CA 91943-2614
Phone: 619-574-7763
Fax: 619-574-1314

Well Mind Association
P.O. Box 201
Kensington, MD 20895-0201
Phone: 301-949-8282
Fax: 301-946-1402

Share, Care and Prayer, Inc.
P.O. Box 2080
Frazier Park, CA 93225

Yeast Consulting Services
Marjorie Crandall, Ph.D., Founder
P.O. Box 11157
Torrance, CA 90510-1157
Phone: 310-375-1073
Fax: 310-791-1363

More About Support Groups

Many communities have support groups formed by people interested in nutrition, mental illness and preventive medicine. And a few communities have support groups for people interested in yeast-related disorders.

Why are support groups needed? From phone calls and letters I've received, I've found that people with yeast-related problems, food and chemical sensitivities and other chronic health problems like to talk to others with similar problems. Yet, they often experience difficulty in locating a group. If that's your situation, you may want to start your own group. Here are suggestions:

- Check with the health editor of your newspaper and/or the public relations director of your hospital.
- Advertise in local papers. In this way you can see if others in your area would like such a group.
- Do networking. Ask members of the staff of your local health food store or pharmacy to help you. Many stores have bulletin boards where you can post notices.
- Pass the word around to church groups or social groups.
- If you know empathetic physicians, nutritionists or other professionals, ask them to help you spread the word.
- Once you get a group of 10 or 12 people, set a time and place for a meeting.
- At the meeting select a group leader/secretary, treasurer and organize your next meeting. Decide from those who come to the meeting the type of help/discussion you want at each meeting.

- Ask a knowledgeable professional to talk to your group.
- Rotate the monthly meeting places. You might want to use various members' homes.

Newsletters and Magazines

Alive—Canadian Journal of Health and Nutrition, published 11 times a year. 7436 Fraser Park Dr., Burnaby, B.C., Canada, V5J 5B9. $24.50 (Canada).

Allergy Alert, Dr. Sally Rockwell, Editor, P.O. Box 31065, Seattle, WA 98103 (206-547-1814). $15/6 issues a year. Also available from this source are books, tapes and telephone counseling.

Allergy Hotline, published monthly by Hotline Printing and Publishing, P.O. Box 161132, Altamont Springs, FL 32716. $35/year.

Alternative and Complementary Medicine, The Journal of, published quarterly by Mary Ann Liebert, Inc., 2 Madison Ave., Larchmont, NY 10538 (914-834-3688).

Better Nutrition for Today's Living, published monthly by Communication Channels, Inc., 6151 Powers Ferry Rd., N.W., Atlanta, GA 30339. Available in most health food stores (404-955-2500).

CanaryNews—Newsletter of the Chicago-area Environmental Illness/ Multiple Chemical Sensitivities (EI/MCS) Support Group, 1404 Judson Ave., Evanston, IL 60201. $15/year.

Clinical Pearls News—A Health Letter on Current Research in Nutrition and Preventive Medicine published by I.T. Services, 3301 Alta Arden #2, Sacramento, CA 95825, (916-483-1085), $109/year, 12 issues.

Delicious Magazine, 1301 Spruce St., Boulder, CO 80302 (303-939-8440). $24/year. Published monthly.

Earl Mindell's Joy of Health, published monthly by Phillips Publishing, Inc., 7811 Montrose Rd., Potomac, MD 20854. $69/year.

Fine Print, a publication of the Hyperactivity Healthline, Inc., P.O. Box 10085, Alexandria, VA 22310. Published quarterly. $20/year.

FM/FMA News Update—A resource guide for those with chronic illness. P.O. Box 8119, Minneapolis, MN 55408-0119. 6 issues per year. Subscription is available only with FM/FMA membership, $35 annually.

Health Confidential, Board Room, Inc., 55 Railroad Ave., Greenwich, CT 06830. Published monthly. $49/year.

Health Hunter, a publication of the Center for the Improvement of Human Functioning International, Inc., 3100 N. Hillside Ave., Wichita, KS 67219. Published 10 times yearly. $25/year.

Health Naturally—Canada's Self-Help Care magazine, Box 580, Parry Sound, Ontario, Canada, PZA 2X5. $19.26 (Canada), $27.00 (U.S.), 6 issues per year (705-746-7809). Fax: 705-746-7893.

"Here's to Your Health!", P.O. Box 130133, Tampa, FL 33681. Fax: 813-282-1132. Published monthly. $29.95 annually.

Dr. Julian Whitaker's Health & Healing—Tomorrow's Medicine Today, published monthly by Phillips Publishing Company, Inc., 7811 Montrose Rd., Potomac, MD 20854 (301-340-2100). $69/year.

Latitudes, Exploring attention disorders, autism, hyperactivity, learning disabilities and Tourette syndrome, 1120 Royal Palm Beach Blvd., #283, Royal Palm Beach, FL 33411. $24/year, 6 issues (407-798-0472).

Let's Live—America's foremost health and preventive magazine. 444 N. Larchmont Blvd., Los Angeles, CA 90004. Published monthly. $19.95/year.

Network News, National Women's Health Network, 514 Kent St., N.W., #400, Washington, DC 20004. Published bimonthly. All members receive a subscription to this newsletter. Membership $25.

NOHA News, published by Nutrition for Optimal Health (NOHA), P.O. Box 380, Winnetka, IL 60093. $8/year, published quarterly.

Nutrition and Healing, a monthly newsletter by Jonathan V. Wright, M.D., Publishers Management Corp., P.O. Box 84909, Phoenix, AZ 85071 (800-528-0559). $49/year.

Sally's Living Without—A Lifestyle Guide to Achieving Better Health, P.O. Box 132, Clarendon Hills, IL 60514-0132. $16/year/4 issues (630-415-3378).

Townsend Letters for Doctors and Patients, 911 Tyler St., Port Townsend, WA 98368-6541 (360-385-6021). $49/10 issues yearly.

The Well Mind Association of Greater Washington, Inc., 1141 Georgia Ave., Suite 326, Wheaton, MD 20902 (301-949-8282) $25/year—includes membership in association.

Women's Health Connection, P.O. Box 6338, Madison, WI 53716-0338, published 6 times a year, $12 (800-366-6632).

Books and Booklets

Albert, R., *Cooking with Rachel**

Ali, M., *The Canary and Chronic Fatigue*

Appleton, N., *Lick the Sugar Habit**

Appleton, N., *Healthy Bones—What You Should Know About Osteoporosis**

Baker, S. M., *Detoxification: Cleansing the Body of the Poisons We Take In—and Those We Create*

Baker, S. M., *Folic Acid*

Balch, J., Balch, P., *Prescription for Dietary Wellness**

Balch, J., Balch, P., *Prescription for Nutritional Healing—A-Z Guide to Supplements** (revised and expanded 2nd edition)

Ballweg, M. L., *The Endometriosis Source Book*

Barnes, B., Galton, L., *Hypothyroidism—The Unsuspected Illness**

Benson, H., *Relaxation Response**
Berthold-Bond, A., *Clean and Green—Complete Guide to Nontoxic and Environmentally Safe Housekeeping**
Berthold, Bond, A., *Green Kitchen Handbook**
Beutler, J., *Flax for Life—101 Delicious Recipes and Tips**
Bland, J., *Digestive Enzymes**
Bland, J., *Bioflavenoids***
Bland, J. *Intestinal Toxicity and Inner Cleansing**
Blaylock, R., *Excicotoxins, The Taste That Kills**
Block, M. A., *No More Ritalin—Treating ADHD Without Drugs**
Borysenko, J., *Minding the Body, Mending the Mind**
Braly, J., *Dr. Braly's Food Allergy and Nutrition Revolution**
Brecher, H., & A., *Forty-something Forever**
Bricklin, M., *Nutrition Advisor**
Bricklin, M. and editors of *Prevention, Practical Encyclopedia of Natural Healing**
Brody, J., *Jane Brody's Good Food Book**
Brown, D. J., *Herbal Prescriptions for Better Health**
Buchman, D. D., *Herbal Medicine**
Canfield, J., Hansen, M.V., *Chicken Soup for the Soul**
Carlson, R., *Don't Sweat the Small Stuff.*
Carper, J., *Miracle Cures**
Carper, J., *Stop Aging Now**
Carper, J., *Food Pharmacy*
Chaitou, L., *Candida albicans—Could Yeast Be Your Problem?**
Challem, J., Dolby, V., *Homocysteine—The New "Cholesterol"**
Challem, J., Dolby, V., *The Health Benefits of Soy**
Challem, J., *Getting the Most Out of Your Vitamins and Minerals**
Chopra, D., *Ageless Body, Timeless Mind**
Chopra, D., *Perfect Health—Mind/Body Program for a Total Well Being**
Cichoke, A., *Enzymes—Nature's Energizer**
Cole, C. L., *Not Milk, Nutmilks!**
Collinge, W., *Alternative Medicine,** American Holistic Health Association Guide*
Conners, C. K., *Feeding the Brain—How Foods Affect Our Children*
Constans, *Great American Smoothies**
Cranton, E., Fryer, W., *Resetting the Clock**
Cranton, E. M., *Bypassing Bypass**
Crayhon, R., *Robert Crayhon's Nutrition Made Simple**
Crook, W. G. and Jones, M., *The Yeast Connection Cookbook*
Crook, W. G., *The Yeast Connection and the Woman*
Crook, W. G., *Chronic Fatigue and the Yeast Connection*
DeMarco, C., *Take Charge of Your Body****
Dadd, D. L., *Nontoxic Home and Office**

Davis, G., *So, Now What Do I Eat?**
Dean, C., *Menopause Naturally**
Dossey, L., *Re-inventing Medicine*
Fenster, C., *Wheat-Free Recipes and Menus: Delicious ïning Without Wheat or Gluten*
Fenster, C., *Special Diet Solutions: Helathy Cooking without wheat, gluten, dairy, eggs, yeast or refined sugar.*
*Gaby, A. R., Magnesium**
Gaby, A. R., *Preventing and Reversing Osteoporosis**
Galland, L., *Power Healing*
Galland, L., *SuperImmunity for Kids**
Gates, D., Bonvie, L., Bonvie, B., *Stevia's Story**
Gates, D., *The Body Ecology Diet**
Gerras, C., *Rodale's Basic Natural Foods*
Gittleman, A. L., *Get the Sugar Out**
Gittleman, A. L., *Natural Healing for Parasites**
Gittleman, A. L., *Guess What Came to Dinner—Parasites and Your Health**
Gittleman, A. L., *The Living Beauty De-Tox Program*
Golan, R., *Optimal Wellness*
Goldbeck, N. & D., *American Whole Foods Cuisine**,**
Goldberg, B. *Chronic Fatigue, Fibromyalgia and Environmental Illness*
*Goleman, D., Gurin, J., Mind Body Medicine**
Golos, N., *If This Is Tuesday, It Must Be Chicken—How to Rotate Food for Better Health**
Griffin, G., Castelli, I., *Good Fat, Bad Fat**
Gursche, S., *Healing with Herbal Juices**,***
Haas, E. M., *The Detox Diet**
Haas, E. M., *The Staying Healthy Shopper's Guide*
Hansen, *Grapeseed Extract—Procyanidolic Oligomers (PCO)**
Hawken, C. M., *St. John's Wort**
Heimlich, J., *What Your Doctor Won't Tell You—Alternative Therapies**
Heinerman, J., *Nature's Super Seven Medicines**
Heinerman, J., *Healing Benefits of Garlic**
Hobbs, C. and Haas, E., *Vitamins for Dummies*
Hoffman, R., *The Natural Approach to Attention Deficit Disorder** (Good Health Guide)
Hoffman, R. L., *Seven Weeks to a Settled Stomach**
Hoffman, R. L., *Lyme Disease**
Hoffer, A., *Hoffer's Laws of Natural Nutrition*
Hunt, G., Bliznakov, E. G., *Miracle Nutrient—CoEnzyme Q_{10}**
Ivker, R. S., *Sinus Survival* (revised), *the Holistic Medical Treatment for Allergies, Asthma, Bronchitis, Colds and Sinusitis**
Jacobson, M. F., Maxwell, B., *What Are We Feeding Our Kids?*

Jensen, B., *Bee Well, Bee Wise with Bee Pollen, Propolis, Royal Jelly**
Johnson, L., Johnston, J. R., *Flax Seed (Linseed) Oil and Power of Omega 3**
Kilham, C., OPC, *The Miracle Antioxidant** (Good Health Guide)
Kinderlehrer, J., *Smart Breakfast**
Krohn, J., Taylor, S. A. and Larson, E. M., *Allergy Relief and Prevention*
Lahoz, S. C., *Conquering Yeast Infections—The Non-Drug Solution*
Langer, S., Scheer, J., *Solve the Riddle of Illness**
Lark, S., *Estrogen Decision**
Lau, B., *Garlic in You—The Modern Medicine**
Lawson, L., *Staying Well in a Toxic World*
Lee, W., *CoEnzyme Q_{10}—Is It Our New Fountain of Youth?**
Lee, W. H., *Friendly Bacteria**
Lieberman, S., Bruning, N., *Real Vitamin and Mineral Book**—2nd Edition
Magaziner, A., *The Complete Idiot's Guide to Living Longer and Healthier*
Marie-Martin, J., and Rona, Z., *Complete Candida Yeasts Guidebook**,***
McDougall, J. & M., *The McDougall Program—Twelve Days to Dynamic Health**
McDougall, J. & M., *McDougall's Medicine—Challenging Second Opinion*
McKeith, G., *Miracle Superfood: Wild Bluegreen Algae**
Miller, B., *PMS—Premenstrual Syndrome**
Mindell, E., *Earl Mindell's Secret Remedies,* and *Earl Mindell's Vitamin Battle,** Revised (1997)
Moeller, M., *Fibromyalgia Cookbook*
Moeller, M. and Elrod, J. M., *The Fibromyalgia Nutrition Guide*
Moraldo, P. and the People's Medical Society, *Women's Health for Dummies*
Mowrey, D. B., *Scientific Validation of Herbal Medicine**
Murray, F., *Ginkgo Biloba—The Amazing 200 Million Year Old Healer**
Murray, M., *Premenstrual Syndrome**
Murray, M., *Encyclopedia of Natural Medicine**
Murray, M. and Pizzorno, J., *Encyclopedia of Natural Medicine*—Revised Second Edition
Murray, M., *Natural Alternatives to Over-the-Counter and Prescription Drugs**
Murray, M., *Natural Alternatives to Prozac**
Murray, M., *Encyclopedia of Nutritional Supplements**
Murray, M., *Chronic Fatigue Syndrome**
Murray, M., *The Healing Power of Herbs*
Null, G., *Nutrition and the Mind**
Null, G., *Complete Guide to Health and Nutrition**
Passwater, R., *Beginner's Introduction to Vitamins**
Passwater, R., *Candaswanic Pycnogenol, the Super Protector Nutrient**

Passwater, R., *Evening Primrose Oil—Its Amazing Nutrients**
Passwater, R., *New Super-Nutrition, Your Guide to Super Vitality and Health**
Pelt, R., *Mind Food and Smart Pills**
Perlmutter, D., *Life Guide—A Guide to a Longer and Healthier Life***
Podell, R. N., *When Your Doctor Doesn't Know Best*
Pritikin, R., *New Pritikin Program**
Quillin, P., *Healing Nutrients**
Quillin, P., Quillin N., *Beating Cancer with Nutrition*
Rajhathy, J., *Free to Fly—A Journey Toward Wellness****
Rapp, D., *Is This Your Child**
Rapp, D., *Is This Your Child's World? How to Fix the Schools and Homes That Are Making Your Child Sick*
Remington, D. W., Higa, B., *Back to Health: Yeast Control**
Richard, D., *Stevia Rebaudian–Nature's Sweet Secret**
Rivera, Rudy, M.D., Deutsch, Roger D., *Your Hidden Food Allergies Are Making You Fat*
Robbins, J., *Reclaiming Our Health**
Robbins, J., *Diet for a New America**
Robbins, J., *Reclaiming Our Health**
Rockwell, S., *Coping with Candida Cookbook**
Rodriguez, L., *Children's Health Problems and Solutions**
Rona, Z. P., *Return to the Joy of Health*,****
Rosenbaum, M., Susser, M., *Solving the Puzzle of Chronic Fatigue Syndrome**
Rosenfeld, *Dr. Rosenfeld's Guide to Alternative Medicine**
Rosenvold, L., *Can a Gluten-Free Diet Help? How?**
Rudin, D., Felix, C., *Omega 3 Oils—To Improve Mental Health, Fight Degenerative Diseases and Extend Your Life**
Russell-Manning, B., *Home Remedies for Candida**
Sahelian, R., *Glucosamine—Nature's Arthritis Remedy**
Sahelian, R., *St. John's Wort—Nature's Feel-Good Herb**
Sahley, B. J., *Healing with Amino Acids**
Sahley, B. J., *Control Hyperactivity/ADD Naturally**—3rd Edition
Schmidt, M., *Healing Childhood Ear Infections**
Schmidt, M., Smith, L., and Sehnert, K., *Beyond Antibiotics**
Schmidt, M., *Tired of Being Tired*
Schmidt, M. A., *Smart Fats*
Semon, B. and Kornblum, L., *Feast Without Yeast*
Sharamon, S., Baginski, B., *Healing Power of Grapefruit Seed**
Shealy, N., *Complete Family Guide to Alternative Medicine**
Shealy, N., *DHEA—The Youth and Health Hormone**
Siguel, E. N., *Essential Fatty Acids in Health and Disease*

Sinatra, S., T., *Heartbreak and Heart Disease—A Mind Body Prescription for Healing the Heart*
Smith, L., *Feed Your Body Right**
Snyderman, N., *Dr. Nancy Snyderman's Guide to Good Health for Women Over Forty**
Sosin, A. and Jocobs, B. L., *Alpha Lapoic Acid*
Spreen, A., *Nutritionally Incorrect, Why the American Diet Is Dangerous and How to Defend Yourself*
Teitelbaum, J., *From Fatigued to Fantastic!*
*The New Our Bodies, Our Selves,** revised, expanded (Boston Women's Health Book Collection)
Thrash, A., Thrash, C., *Natural Remedies—A Manual**
Thrash, A., Thrash, C., *Natural Healthcare for Your Child**
Thrash, A., and Thrash, C., *Food Allergies Made Simple*
Trenev, Natasha, Probiotics: Nature's Internal Healers
Truss, C. O., *The Missing Diagnosis**
Walker, M., *Olive Leaf Extract*
Weil, *Natural Health—Natural Medicine**
Weil, A., *Healthy Living**
Weil, A., *Health and Healing**—revised, updated
Weil, A., *Common Illnesses**
Weinberger, S., *Candida albicans, A Quiet Epidemic**
Weiner, M., *Healing Children Naturally**
Weintraub, S., *Natural Treatment for ADD and Hyperactivity**
Werbach, M. R. and Murray, M., *Botannical Influences on Illness**
Whitaker, J., *Dr. Whitaker's Guide to Natural Healing**
Winderlin, C. and Sehnert, K., *Candida Related Complex*
Wright, J., *Dr. Wright's Guide to Healing with Nutrition**
Wright, J., Morgenthaler, J., *Natural Hormone Replacement**
Wunderlich, R., *Natural Alternatives to Antibiotics**
Wunderlich, *The Candida Yeast Syndrome**
Yutsis, P., Gelsky, M., *The Candidiasis Revolution*
Yutsis, P., Walker, M., *The Downhill Syndrome—If Nothing's Wrong, Why Do I Feel So Bad?*
Zand, J., Walton, R., Rountree, B., *Smart Medicines for a Healthier Child**
Ziff, S., *Silver Dental Fillings—The Toxic Time Bomb**
Zukin, J., *Raising Your Child Without Milk**

---

*Books available in your health food store.
**Books or booklets $10 or less.
***Available in Canada.

APPENDIX · C

Postscript

In acquiring materials for this book, I read dozens of articles in scientific journals and lay publications and consulted many knowledgeable professionals and nonprofessionals. I also learned from the observations of people who called and wrote me. When I completed the manuscript of the first edition of this book and sent it to my typesetter, John Adams, he said, *"No more changes!"*

But, in the next two years, I received so much new information relevant to people with yeast problems that I'm including it in this 1999 revision.

More About Oral Antifungal Medications

Nonprescription Agents for Yeast Infections

Probiotics

In a comprehensive seven-page article, researchers from the School of Medicine, University of Washington, Seattle, reviewed the medical literature and described their own observations on the importance of probiotics, including *Lactobacillus acidophilus* and *Bifidobacterium bifidum*. In the conclusion of their article, they said,

> There is now evidence that administration of selected microorganisms is beneficial in the prevention and treatment of certain intestinal and, possibly treatment of vaginal infections. *In an effort to decrease the reliance on antimicrobials, the time has come to carefully explore the therapeutic applications of biotherapeutic*

agents. (emphasis added) (Elmer, G. W., Surawicz, C. M. and McFarland, L. V., "Biotherapeutic Agents—A Neglected Modality for the Treatment and Prevention of Selected Intestinal and Vaginal Infections," *JAMA,* March 20, 1996, 275:870–876)

Oregano

I first heard about this perennial pungent herb of the mint family from Ann Fisk, R.N., a member of the Advisory Board of the International Health Foundation. A short time later, Jeffrey S. Bland, Ph.D., discussed oregano in the September 1996 issue of *Preventive Medicine Update.* I obtained further information in an article by Robert A. Ronzio, Ph.D., and colleagues.

Based on the scientific studies presented in this article, these researchers concluded that oregano is an effective antiyeast agent and more potent than caprylic acid. (Stiles, J. C., Sparks, W. and Ronzio, R. A., "The Inhibition of *Candida albicans* by Oregano," *J. Applied Nutrition,* 1995; 47:96–101.)

I've also received favorable reports from several health professionals who are using oregano in treating patients with yeast-related disorders. To obtain additional information I consulted Jonathan Wright, M.D., Kent, Washington. He told me that he had found oregano to be a safe and excellent antiyeast agent. Moreover, he said that in his experience it was as effective as nystatin. His usual dose: one 50 mg. tablet four times daily.

Information about a special emulsified form of oregano (A.D.P.) can be obtained from Biotics Research, P.O. Box 36888, Houston, TX 77236. (FAX 281-240-2304, 1-800-231-5777.)

Undecylenic Acid

Fatty acids have been known and used for centuries as antimicrobial agents, originally in the manufacture of soaps. In the last 50 years, however, they have found use both *invitro* as yeast and mold inhibitors in food stuffs and as topical, intestinal and systemic antifungals.

One of these fatty acid products, caprylic acid, has been used successfully by many health professionals and their patients during the last several decades (see pages 129–130). Recently, I received information about another fatty acid product, undecylenic acid (UA), an 11-carbon mono-unsaturated fatty acids which is produced commercially by the vacuum distillation of castor bean oil.

According to a 1997 article in *Alternative Medicine Review (AMR)* (Volume 2, No. 5, page 350, 1997), UA has been used as a topical antifungal agent in over-the-counter products including Desenex. It has also been used in humans orally as an economical antifungal agent. As much as 20 grams of UA have been taken by mouth and caused no toxicity and only mild gastrointestinal symptoms.

I learned more about UA from Kathy Gibbons, Ph.D. and Susanna Choi, M.D., who use it in treating their patients with yeast-related disorders. Kathy told me they'd found it to be especially helpful following a course of Diflucan or Sporanox.

She also said that some of their patients found UA works just as well as the prescription medications. Usual dose: three to five capsules three times a day. To obtain more information about this product which is marketed under the name, "Formula SF 722," contact Thorne Research, Inc., Phone 208-263-1337, Fax 208-265-2488. Although it can be ordered from the pharmacies listed on page 121, I feel that its use should be supervised by a knowledgeable health professional.

Bigger Doses of Nystatin May be Needed

In August 1996, during a phone conversation, candida pioneer, C. Orian Truss, M.D., again stressed the value and safety of nystatin in treating patients with yeast-related disorders. Yet, he pointed out that much larger than the usual recommended doses may be needed to curb candida overgrowth in some patients—especially in those with weakened immune systems.

Dr. Truss said in his experience, *Candida albicans* does not become resistant to nystatin. He cited an article published 15 years ago which described the response of children with severe weakness of the immune system to larger doses of nystatin. (*New England Journal of Medicine,* Vol. 304: April 30, 1981; pp. 1057–62)

So here's my "bottom line" advice based on the experiences of Dr. Truss and other researchers and clinicians: "If your health problems are yeast connected and have not responded to nystatin, don't give up until you've taken dosages in the range of four to eight million units, four times a day."

Lamisil—Another Antifungal Medication

According to information I received from the Sandoz Pharmaceutical Company, "Lamisil (terbinafine hydrochloride . . .) is a member of a new class of antifungals, the allylamines, which have a different mechanism of action from the classic azoles. It selectively inhibits an early step of the fungal cell membrane sterol synthesis, which both interferes with the synthesis of the fungal cell membrane and causes an accumulation of a toxic precursor molecule (squalene) in the organism."

In 1995, Sandoz received approval from the FDA for the use of Lamisil tablet (oral) in the treatment of fungal infections of the toenails and fingernails due to *Trichophyton rubrum* and *Trichophyton mentagrophytes.*

It is dispensed in 250 mg. tablets, and according to the package insert, in doses of one tablet a day most strains of dermatophyte fungi when causing nail infections are susceptible to this drug. Clinical success rates for candida infections of the nail at the 250 mg. daily dose, six weeks for fingernails and 12 weeks for toenails, are still being evaluated in controlled trials.

My Comments

After receiving this information, I called a physician who has successfully treated hundreds of yeast patients using Diflucan, Sporanox and dietary changes. He told me that he had received samples of Lamisil two months ago and was using it in treating six of his patients with yeast-related disorders. He said,

> I've found it to be a useful adjunct in treating these patients and I noted no adverse reactions.

Lamisil has received approval by the FDA only for use in treating dermatophyte fungal infections of the nail. So the situation with this drug resembles that of Sporanox which has received the same limited approval. Like all antifungal medications, Lamisil may cause side effects. Accordingly, it will take further scientific study and clinical experience to determine the place of Lamisil in treating patients with a broad group of chronic health disorders.

Diflucan Is Safe for Young Children

According to a July 18, 1996 letter I received from John D. Ostrosky, Pharm.D., Associate Director of Medical Information of the Pfizer Pharmaceutical Company,

> Diflucan has shown to be safe and effective in children six months to 13 years old. Pediatric patients have been successfully treated with doses of 3mg/kg/day to 12mg/kg/day.

This data was gathered from several studies using Diflucan in immunocompromised children with candida of the mouth and throat. In one of these studies, the clinical and mycological response rates of the children treated with Diflucan were higher than in a second group treated with nystatin. In addition, the incidence of adverse side effects in both groups of patients were minimal and were no greater in the Diflucan group than in the nystatin group.

My Comments

At this time (late summer, 1996) the FDA has not given approval for the use of oral antifungal medications in treating a diverse group of health disorders that affect people of all ages and both sexes, including children with ADHD. Yet, based on the reports cited by Ostrosky, and my own experiences in practice, Diflucan in a dose of 50–100mg/day in children with ADHD who have received repeated antibiotic drugs is a treatment option worth considering.

I'm not pleased, however, with the liquid preparation of Diflucan which is now available for use in infants and young children. Here's why: Like the liquid preparations of nystatin, it contains sugar. (See also pages 62 and 122.)

Antifungal Medication for Depression

In May 1996, two Boston psychiatrists presented case history evidence that ketoconazole (Nizoral) is effective in chronic atypical depression. And they presented their observations on two patients seen in their practice.

The first of these patients was a 44-year-old woman with a 7-year history of persistent depressive symptoms and a second patient was a 35-year-old woman with a history of chronic atypical depression since childhood. In each of these patients the depression and associated symptoms, including decreased libido and fatigue, responded favorably to the administration of ketoconazole (Nizoral).

In their concluding paragraph, these observers said,

> Our cases suggest that the antidepressant response spectrum for ketoconazole might extend to chronic atypical depression and that the drug may be useful to those patients who cannot tolerate antidepressant side effects. . . . Double blind studies of this drug in chronic atypical depression are worth pursuing. (Sovner, R. and Fogelman, S., *Journal of Clinical Psychiatry,* Vol. 57:5, pp. 227–228, May 1996)

More About Children

Essential Fatty Acids Help the Brain

If your automobile . . .
. . . won't start
. . . gets only six miles to the gallon
. . . sputters, shakes and knocks
. . . puts out blue smoke,

. . . wouldn't you check on the type and quality of fuel you're putting in your car's gas tank?

Shouldn't we also look at the "fuel" being put in children's "gas tanks"?

My answer to these questions during the past 25 years has been a loud "YES."

So I was excited to read an article from Purdue researchers in the spring of 1996 which provides scientific support for nutrients often lacking in children's diets—*essential fatty acids*. In their report Laura J. Stevens and colleagues described their research observations. Here's an excerpt from an abstract of their study:

> A greater number of behavior problems, assessed by the Conners' Rating Scale, temper tantrums, and sleep problems were reported in subjects with the lower total Omega-3 fatty acid concentrations. Additionally, more learning and health problems were found in subjects with lower total Omega-3 fatty acid concentrations. (Stevens, L. J. et al., "Omega 3 Fatty Acids in Boys with Behavior, Learning and Health Problems," *Physiol Behav,* (4/5)915–920, 1996)

Based on this early observation and other scientific observations cited by Stevens and her colleagues, parents should make sure that their children obtain adequate amounts of these important nutrients. (See also pages 99–100.)

Milk and Wheat May Cause Brain Dysfunction

At a June 1996 conference of the Developmental Delay Registry (DDR),* in Orlando, Florida, J. Robert Cade, M.D., of the University of Florida, discussed his research findings on people with autism and schizophrenia. In his presentation he pointed out that in both of these disorders there is an increase in the level of peptides in the urine. These peptides come from casein in milk and other dairy products and from gliadin and gluten in wheat, barley, oats and rye.

On a gluten/casein free diet, 81% of a group of autistic children improved within 3 months. Symptoms and signs in these children which improved include, social integration, eye contact, hyperactivity and panic attacks. Yet, Dr. Cade pointed out that although autistic children made significant improvement in all categories, a single breakdown in diet may cause a relapse that will take three months for the symptoms to return to normal.

For more information write to J. Robert Cade, M.D., Depts. of Med. & Physiol., P.O. Box 100204, University of Florida, Gainesville, FL 32610.

---

*DDR, 6701 Fairfax Rd., Chevy Chase, MD 20815.

Dietary Factors in ADHD Continue to Be Overlooked by Many Professionals*

A May 1996 article in the official newspaper of the American Academy of Pediatrics *(AAP News),* cited a survey of 380 pediatricians who said that parents' belief that poor diet causes ADHD was a "misperception."

This report prompted me to respond and my comments were published in the July 1996 issue of *AAP News* under the heading, "ADHD: Dietary Factors Overlooked." I pointed out that during my years of pediatric practice many of my patients with chronic health problems, including ADHD, improved following dietary changes. Moreover, I reported my observations in major medical journals on a number of occasions. In my concluding comments I said,

> Based on my own experience and scientific reports in the medical literature, I'm unable to accept the recommendations of those who appear to suggest that millions of American children are suffering from what might be termed "a Ritalin deficiency . . ."
>
> I urge the AAP Committee on Children with Disabilities and other appropriate AAP committees to host a "workshop/think tank" on ADHD, its causes and management. Participants should include pediatricians and other professionals and nonprofessionals with different experiences and persuasions. The lives of countless children and their families are at stake.

New Help in the Computer Age

For Parents of Autistic Children

At a workshop/think tank on autism, I met Lisa S. Lewis, Ph.D., of New Jersey. She is the parent of an autistic child; occupation: anthropology/computer science. Her E-Mail address: lisas@pucc.princeton.edu.

In August 1996 I called Lisa and asked her for additional information. She responded with three pages of information with the heading, "Surfing the Net for Fun and Information." Here are excerpts:

---

*See also page 82.

> With a computer, with a (relatively high speed) modem (software), you have a wealth of information at your fingertips. With these three things you will have access to the Internet, the so-called "information superhighway," as well as to electronic mail and news groups on every subject imaginable . . .
>
> There are many lists about topics of yeast, autism, ADD/ADHD and other disabilities. Your software will allow you to look at the names of all available groups and choose those you wish to view regularly . . .
>
> The amount of information that is available, once you have the proper hardware and software, is nearly limitless. It increases daily and so far there's no end in sight. Perhaps as important as what you will learn, is how many people there are out there in the same boat. There is much to learn, and much to share. Good luck and happy surfing!

Included in Lisa's information were pages about autism, pages about food intolerance and allergy and pages about alternative medicine, yeast and homeopathy.

For Children with ADHD

In you have access to the Internet and are searching for more information about children with ADD/ADHD, then you'll want to visit Laura Steven's home page at: http://www.222.nlci.com/nutrition.

This "newsletter" brings you the latest information on the known biochemical factors that cause ADHD, including food sensitivities, lead toxicity, thyroid problems, etc.—and what to do about them. You can select different topics of interest: "News You Can Use," "Recipe of the Month," "Medical References for You and Your Doctor," "Questions and Answers," and "Comments from Readers." Each month Laura updates the home page. And if you have any questions, you can E-mail her at: lstevens@pop.nlci.com. You'll like this user-friendly newsletter.

For People with Food Allergies

If your health problems are yeast-connected, you're probably bothered by food allergies and sensitivities. As I noted on pages 214–215 of this book, by rotating your diet you can better manage and overcome them. And in June 1996, I obtained information which could make your job easier.

I'm referring to the Rotation Diet Planner Software, a computer program designed by Registered Dietician, Bettina Newman, of Chesapeake, Virginia. Bettina developed this program in 1993 in response to the need expressed by her patients with candida and food allergies/sensitivities for very personalized rotation diets.

The simple and fun-to-use point and click Windows technology allows the user to select favorite foods from a data base of more than 800 foods and then to place them in preferred combinations for menu planning.

To obtain information call (804) 547–5190, or E-Mail requests to newman@pinn.net for a free demo disk ($2.50 s/h), to place an order or if you have any questions.

Yeast Infections in Other Parts of the Body

In my discussion of yeast problems in my publications and lectures, I always point out that *Candida albicans* thrives on the warm *interior* membranes of the body, especially the digestive tract and vagina. And I've paid little attention to candida infections of the skin and mouth.

In September 1996, I received a Fax from Marjorie Hurt Jones, R.N., co-author of *The Yeast Connection Cookbook,* which made me realize I had neglected several common yeast problems. Here's what Marge had to say:

> I understand that your main focus in *The Yeast Connection Handbook* and your other publications is on the gut and resulting manifestations which make people feel "sick all over." Yet, I'm sure that many people have yeast infections elsewhere, including under the arms and breasts, in the groin and in the mouth. So I hope that in your discussion of yeast problems in your usual comprehensive manner, in *The Yeast Connection Handbook* you'll include these areas too.

Yeast Infections of the Skin

To emphasize the importance of yeast infections of the skin, Marge described her own recent experience.

> During a 90+ degree heat wave this summer, I developed yeast infections under my arms. And for a few weeks, I smelled like rising bread! No joke. For about ten days I tried treating it with local antiyeast powders and ointments I found in the drug store. Although this treatment helped a little, it didn't get the job done. So I called my physician, who prescribed Diflucan by mouth, which promptly cleared it up. My experience leads me to say that yeast infections of the skin may not always be a self-help situation and a person may need a prescription antifungal to treat them.

Yeast Infections of the Mouth (thrush)

Thrush, a mild yeast infection, is sometimes seen in healthy infants during the first few days or weeks of life. It may develop because infants are exposed to yeasts as they pass through the birth canal during vaginal delivery. Thrush looks as if patches of milk are stuck to the cheeks, tongue and roof of the mouth. But unlike milk, it does not wipe off easily. Happily most infants respond promptly to simple local treatment measures prescribed by the physician, including oral nystatin or gentian violet.

Yeast infections of the mouth occur also in people of any age who take repeated or prolonged courses of antibiotic drugs and/or people with weakened immune systems. Thrush is especially apt to develop in individuals with dentures. Such infections, like yeast infections elsewhere, often require a sugar-free special diet and oral antifungal medication, including nystatin powder or one of the azole drugs. (For my comments on liquid nystatin see page 122.)

More About the Yeast Connection to Endometriosis

In 1996, Mary Lou Ballweg, Executive Director of the Endometriosis Association (EA), published two 8-page newsletter articles entitled, "Endometriosis and *Candida albicans:* Even More Startling Connections." Here are brief excerpts:

> When the Endometriosis Association was just three years old, several members discovered something amazing: when they were treated for allergic symptoms resulting from a common yeast, *Candida albicans* . . . the endometriosis symptoms *also*

cleared up. We published a short note in our newsletter in January 1984 to see if other members had had related experiences...

The feedback on *Candida albicans* and its expanding universe of related problems hasn't stopped since. In fact, looking back over the association's sixteen-year history, it's fair to say that *no other approach to endometriosis treatment has given as consistent and long term positive results as has treatment for Candida albicans/allergy/infection and its related problems.* (emphasis added)

In Part II of her newsletters, Ballweg included a section entitled, "Members Experiences with Candida." Included was a letter from a woman with repeated vaginal yeast infections, endometriosis, depression, fatigue, PMS, chemical sensitivities and many other symptoms. Here is an excerpt:

Don't give up. If one doctor looks at you like you've come from another planet, go to one who will listen. There is no reason why doctors have to be rude. You know your body . . . Read all you can in order to know what others have learned. It will save you money and a lot of heartache. Remember you're a good person even though you don't understand what is happening to you. There is help out there. Once you begin to help yourself, help others, get involved with such groups as EA.

To obtain information about these newsletters, write to Endometriosis Association, 8585 N. 76th Place, Milwaukee, WI 53223. (See page 69 for more information about this outstanding organization.)

Alternative Medicine and Herbal Remedies

In the two years since the first printing of this book, interest in and support for alternative medicine and herbal remedies have zoomed like a rocket. Here are a few examples: An article in the November 1997 issue of the *Harvard Healthletter (HHL),* entitled, "Alternative Medicine—Time for a Second Opinion," said,

Although alternative medicine may not be so "alternative" anymore, most Americans still don't discuss their use of unconventional therapies with their physicians and many doctors are quick to dismiss interest in them. But, times are changing and some physicians are urging their peers to be more open-minded and to listen to what patients say has helped them.

To find information on herbs, *HHL* told readers to consult books by pharmacologist, Varro Tyler, Ph.D., a recognized expert on plant-based drugs and a former professor and member of the faculty at the Purdue University School of Pharmacy. *HHL* also recommended the American Botanical Council (ABC) headed by my friend, Mark Blumenthal. More information can be found on your web site at www.herbalgram.org or you can write to ABC at P.O. Box 201660, Austin, TX 78720–1660.

In an article in the December 1997 issue of *Let's Live,* Beth Salmon describes her visit with Harvard-trained, Andrew Weil, who is working, "to change the face of mainstream medicine." Here are excerpts,

> Fueled in part by economics, he suggests there may be nothing short of a revolution taking place in health care . . . He's dedicated the past two decades to chipping away what he beholds as the faulty and misguided premises of modern medicine by boldly redefining it. Inspirited in ten books; a monthly newsletter, *Dr. Andrew Weil's Self-Healing;* a popular web site, *Ask Dr. Weil,* which garners about two million hits per months; as well as assorted multimedia products, Dr. Weil's message is reaching the masses.

In May 1998, Dr. Weil sent me information about the work he and his colleagues are doing at the University of Arizona. Here are excerpts from Tracy W. Gaudet, M.D., Medical Director, Program in Integrative Medicine.

> The vision of Dr. Andrew Weil, our director, is one of a new kind of medicine, a medicine that combines the best ideas and practices of conventional medicine with the best ideas and practices of alternative medicine into treatments that stimulate the body's natural healing abilities.
>
> This new medicine requires a different approach to health and healing, a different relationship of the patient . . . a different kind of practitioner . . . Here at the University of Arizona, we've begun the first program designed to train practitioners in this new approach to Integrative Medicine.

I like the term, *integrative,* and it is being used by more people including physicians. The term makes sense because traditional medicine is needed by many people.

More About Treating Severe Chronic Fatigue

Jacob Teitelbaum, M.D.*, and B. Bird published their findings on 64 patients who experienced severe fatigue and other symptoms for an average of three years. In helping their patients, they used a number of interventions, including treatment of yeast overgrowth, thyroid and adrenal dysfunction, viral and parasitic infections and micronutrient deficiencies (*J. of Musculoskeletal Pain,* 3(4): January 1996).

In 1998, Dr. Teitelbaum and colleagues completed a placebo-controlled study of 70 patients with Fibromyalgia/Chronic Fatigue Syndrome using multiple therapies including the antifungal drug, Sporanox. The patients in the treated group enjoyed "significantly greater benefit" when compared to the placebo group ($p < .0001$).

More Nonprescription Agents for Yeast Infections

Olive Leaf Extract: I've received reports of the effectiveness of this natural product from several people who have written to the International Health Foundation. I obtained additional information from Stan Meyerson, N.E.E.D.S., Syracuse, New York (800-634-1380) and Larry Stephens, Wellness Health and Pharmaceuticals, Birmingham, AL (800-227-2627). These pharmacists told me that a number of health professionals had reported on the value of Olive Leaf Extract in treating people with viral infections and yeast-related problems.

Kolorex: I learned about this New Zealand herbal product from Irwin Rosenberg, The Apothecary, Bethesda, Maryland (800-869-9159). This pharmacist sent me information including an article by Arnold Fox, M.D. who said "Kolorex kills Candida more rapidly and effectively than most medicines."

---

*Dr. Teitelbaum has published his observations in a 1996 book, *From Fatigued to Fantastic.* (See page 247.)

Index

IF after reading this book you'd like more information, get a copy of the 768-page *The Yeast Connection and the Woman.* Here are additional comments made by Philip K. Nelson, M.D., in his Introduction:

"I'm so impressed by:
- the author's willingness to tackle such a comprehensive project
- his fair approach in presenting all sides of controversial topic
- his scholarship and research, including literature citations for those who wish to pursue a particular topic
- his ability to utilize a vast network of health professionals who may be expert in fields where he is not
- his careful use of quotes and attributions
- his care in not making exaggerated health claims."

IF you're troubled by Chronic Fatigue Syndrome, you'll find a comprehensive discussion of this often misunderstood disorder in *Chronic Fatigue Syndrome and the Yeast Connection.* In the Foreword of this book, Carol Jessop, M.D., commented:

> This book does not claim that the common yeast, *Candida albicans* is *the* cause of CFS; however, it does explain the role of multiple entities: yeast overgrowth, intestinal parasites, unchecked viral infections, food allergies and chemical sensitivities and how these can result in the immune dysregulation we refer to as CFS.

IF you need more information about what foods you can eat, the foods you need to avoid, and how to prepare them, read Section II (pages 113–359) of *The Yeast Connection Cookbook.* You'll find over 200 recipes which will help you prepare foods that you can eat and enjoy. Here are excerpts from Marge's introduction:

> I've emphasized tasty vegetables of all sorts which will make your diet more enjoyable and less apt to cause allergies. I also use a variety of grains rather than just wheat and corn. And I introduce you to grain alternatives, including various starches, amaranth and quinoa, and later buckwheat . . . I can't deny that cooking takes time, I can only suggest that *if you really want to enjoy better health, planning and preparing nutritious meals is the place to start.*